The Lecture Notes are intended to report quickly and informally, but on a high level, new developments in mathematical economics and operations research. In addition reports and descriptions of interesting methods for practical application are Particularly desirable. The following items are to be published:

1. Preliminary drafts of original papers and monographs

2. Special lectures on a new field, or a classical field from a new point of view

3. Seminar reports

4. Reports from meetings

Out of print manuscripts satisfying the above characterization may also be considered, if they continue to be in demand.

The timeliness of a manuscript is more important than its form, which may be unfinished and preliminary. In certain instances, therefore, proofs may only be outlined, or results may be presented which have been or will also be published elsewhere.

The publication of the *"Lecture Notes"* Series is intended as a service, in that a commercial publisher, Springer-Verlag, makes house publications of mathematical institutes available to mathematicians on an international scale. By advertising them in scientific journals, listing them in catalogs, further by copyrighting and by sending out review copies, an adequate documentation in scientific libraries is made possible.

Manuscripts

Since manuscripts will be reproduced photomechanically, they must be written in clean typewriting. Handwritten formulae are to be filled in with indelible black or red ink. Any corrections should be typed on a separate sheet in the same size and spacing as the manuscript. All corresponding numerals in the text and on the correction sheet should be marked in pencil. Springer-Verlag will then take care of inserting the corrections in their proper places. Should a manuscript or parts thereof have to be retyped, an appropriate indemnification will be paid to the author upon publication of his volume. The authors receive 25 free copies.

Manuscripts written in English, German, or French will be received by Prof. Dr. M. Beckmann, Department of Mathematics, Brown University, Providence, Rhode Island 0 29 12/USA, or Prof. Dr. H. P. Künzi, Institut für Operations Research und elektronische Datenverarbeitung der Universität Zürich, Sumatrastraße 30, 8006 Zürich.

Die Lecture Notes sollen rasch und informell, aber auf hohem Niveau, über neue Entwicklungen der mathematischen Ökonometrie und Unternehmensforschung berichten, wobei insbesondere auch Berichte und Darstellungen der für die praktische Anwendung interessanten Methoden erwünscht sind. Zur Veröffentlichung kommen:

1. Vorläufige Fassungen von Originalarbeiten und Monographien.

2. Spezielle Vorlesungen über ein neues Gebiet oder ein klassisches Gebiet in neuer Betrachtungsweise.

3. Seminarausarbeitungen.

4. Vorträge von Tagungen.

Ferner kommen auch ältere vergriffene spezielle Vorlesungen, Seminare und Berichte in Frage, wenn nach ihnen eine anhaltende Nachfrage besteht.

Die Beiträge dürfen im Interesse einer größeren Aktualität durchaus den Charakter des Unfertigen und Vorläufigen haben. Sie brauchen Beweise unter Umständen nur zu skizzieren und dürfen auch Ergebnisse enthalten, die in ähnlicher Form schon erschienen sind oder später erscheinen sollen.

Die Herausgabe der „*Lecture Notes*" Serie durch den Springer-Verlag stellt eine Dienstleistung an die mathematischen Institute dar, indem der Springer-Verlag für ausreichende Lagerhaltung sorgt und einen großen internationalen Kreis von Interessenten erfassen kann. Durch Anzeigen in Fachzeitschriften, Aufnahme in Kataloge und durch Anmeldung zum Copyright sowie durch die Versendung von Besprechungsexemplaren wird eine lückenlose Dokumentation in den wissenschaftlichen Bibliotheken ermöglicht.

Lecture Notes in Operations Research and Mathematical Economics

Edited by M. Beckmann, Providence and H. P. Künzi, Zürich

2

U. Narayan Bhat

Case Western Reserve University, Cleveland, Ohio

A Study of the Queueing Systems M/G/1 and GI/M/1

1968

Springer-Verlag Berlin Heidelberg GmbH

ISBN 978-3-662-38801-3   ISBN 978-3-662-39706-0 (eBook)
DOI 10.1007/978-3-662-39706-0

Library of Congress Catalog Card Number 68-19088. Title No. 3752.

PREFACE

This study has grown out of a part of the author's thesis "Some Simple and Bulk Queueing Systems: A Study of Their Transient Behavior" submitted to the University of Western Australia (1964) and a course on Queueing Theory given to graduate students in the Operations Research Group of Case Institute of Technology, Cleveland, Ohio. The one semester course (approximately 35 hours) consisted of the following topics.

(i) Some of the important special queues such as M/M/s, M/D/s, $M/E_k/1$ etc., with emphasis on the different methods employed in the transient as well as steady state solution.

(ii) Imbedded Markov chain analysis of M/G/1 and GI/M/1 as given in the joint paper of the author and N. U. Prabhu as well as the papers of D. G. Kendall. [All notations and papers are referred to later in the notes].

(iii) The contents of this memorandum.

The author feels that such a course prepares the students adequately for an advanced course in Queueing Theory involving topics on Waiting Times, the General Queue GI/G/1 and other ramifications such as Priorities, etc.

A few words regarding the approach adopted in this study may not be out of place.

So far, the time dependent behavior of queueing systems has not found a place in courses given outside the Department of Mathematics. In the mathematical analysis of the systems a wide variety of techniques have been developed and used and some of the most complicated systems have been investigated. However, even though queueing systems are systems of real life situations, improvements in the design of such systems do not match with the increase in theoretical developments. There seems to exist a wide gulf between the theoretical researcher and the one who uses his results. It seems to us that this is due to the methods of analysis used and the form of results obtained. Analytically these methods are powerful and can be used in very complex situations as demonstrated in the papers mentioned in bibliographies on the subject. However, there is a serious disadvantage in these methods, which an analytically oriented researcher tries to neglect. Plainly speaking, the results, given in terms of transforms, very often with more than one argument, fail to make sense to an applied researcher. In simpler situations means and variances could be deduced without much difficulty from these transforms. But, as the transforms get complicated, even these operations need extra dexterity in mathematical manipulations. This seems to us, has been the main reason for the gulf that exists between the theoretician and the applied scientist, even though so much ingenuity has been shown in tackling a variety of technical problems on paper by some of the ablest people in the world. As Kendall (1964) has aptly put it "much of the detail of the queue - theoretic scene has been obscured by the Laplacian curtain". As a result, a researcher with not much sophistication in mathematical technqiues, takes the easiest and simplest way out of his complex problems by assuming steady state from the start of the operation and/or saying that all arrival or service time distributions can be approximated by an exponential distribution, without loss of significant information.

Over the last few years several researchers have directed their attention in remedying this defect and for the single server systems with (i) Poisson arrivals and general service times and (ii) general independent arrivals and negative exponential service times, a good number of combinatorial approaches have come out. Some of the investigations of Prabhu (1965), Takács (1967) and the present author belong to this category. Takács uses generalization and extensions of the classical Ballot theorem to give a combinatorial study of the underlying stochastic process. The present author worked closely with Prabhu and has been able to add much to what has been done by Prabhu and straighten out some of his methods which may seem to be roundabout.

The method given by Prabhu is based on the study of the waiting time processes in these systems. It is true that if either one of the waiting time or queue length processes is studied, the other one follows automatically. However, it should also be noted that the queue length process is discrete-valued and independent of the queue discipline to a large extent. Because of this property a study based on the queue length process has several advantages.

The present author has developed such a method which deals with the queue length process directly and makes minimal use of transforms. The only time we use transforms is while getting some steady state results. Although it is possible to derive such results without resorting to transforms, an appeal to transforms makes it much simpler - and it is our aim to achieve simplicity in methods.

With a few exceptions, the results presented in the following pages have appeared elsewhere in journals and books. However, as more and more schools have started teaching Queueing Theory formally, we hope the approach given in this memorandum will be very useful in teaching the time dependent behavior of queueing systems, which seems to be finding its due place in applied research lately.

In this work the author has received direct and indirect encouragement from Professor N. U. Prabhu of Cornell University. Grateful acknowledgments go to him. I would be failing in my duty if I do not thank Miss Susanne Preston for her excellent job in typing this manuscript.

TABLE OF CONTENTS

INTRODUCTION

A 'queue' is a waiting line of units demanding service at a service facility (counter); the unit demanding service is called the 'customer' and the device at which or the person by whom it gets served is known as the 'server'. This terminology is used in a wide context. Here are a few realistic examples of this customer - server mechanism.

(i) Vehicles demanding service arrive in a garage and depending on the number of employees, one or more vehicles may be repaired at a time.

(ii) Patients arrive at a doctor's clinic for treatment. Even if some appointment system exists, due to the emergency service rendered, there is a random element present in the arrival scheme and, there is a possibility of a waiting line building up.

(iii) In a telephone exchange, the incoming calls are the customers who demand service in the form of telephone conversations.

(iv) Passengers demanding tickets queue up in front of a ticket counter.

It is possible to give numerous examples of this type, where a queue situation exists in one form or the other. As suggested by the above problems, some of these situations differ from each other in several details. However, it is not difficult to see that all these situations have certain common basic characteristics.

(i) <u>Input processes</u>: If the arrivals and service are strictly according to schedule a queue can be avoided. But they are not, especially the arrivals. In most situations arrivals are controlled by external factors and these factors contribute to the uncertain nature of arrivals. For instance, the arrivals could be in groups of random or constant size or in the simplest case, one at a time. The time intervals between successive arrivals can be considered as random variables, having certain distributions. Further, the arrivals could be emenating from a finite or an infinite source. For instance, if the customers are the machines needing repair in an industrial concern the number of machines is a finite number; whereas customers at an ordinary store window could be considered as coming from an infinite source. Therefore, the source of arrivals, the type of arrivals and the interarrival times should be specified in a complete specification of an input process.

(ii) <u>Service mechanism</u>: The number of servers is an integral part of the service mechanism. So also are the duration of service and whether service is given to groups of customers or to one at a time. Because of the uncertainty involved in the length of service - a telephone conversation is a good example - we can consider the service time as a random variable having a certain distribution.

(iii) <u>Queue discipline</u>: We can consider all other factors regarding the rules of the conduct of the queue under this heading. The simplest of these is known as the first-come, first-served discipline. This specifies that customers arriving when the server is busy will be taken for service in the order of their arrival. This rule can be changed to others such as 'last-come, first served', 'random selection for service' etc. Clearly when customers arrive in groups, it is assumed that those in a group are ordered for the sake of service. In addition to these one may introduce some sort of variations in customers getting impatient after waiting, may leave the system without getting served. Some may jocket for positions in the waiting line. Or some customers may be considered as having higher priorities in service than others. The system may not allow more than a certain number to be waiting at a certain time; that is, the size of the waiting room could be finite, and those arriving when the waiting room is full are allowed to be lost to the system. These are only a few examples of various types that can be derived from the simple system described. It is essential, therefore, for the complete description of a queueing system, the rule to be followed has to be specified.

Based on these descriptions the problems arising in Queueing Theory can be classified into three.

(1) <u>Behavioral problems of the system</u>. The study of behaviroal
problems aims at understanding a particular situation as thoroughly as
possible. This is done by using mathematical models. Naturally, these
are idealized models to varied degrees of realism. As done in many other
branches of science, these models are studied analytically in isolation,
hoping that the information obtained from such a study would be useful in
the decision making process regarding such situations. Some of the
characteristics considered in this connection are the distributions, expected
values and other moments connected with the queue length, waiting time and
the length of the busy period in a queueing system.

(2) <u>Statistical problems of the system</u>. By statistical probelms we
mean the problems of the study of empirical data, estimation and tests of
hypotheses regarding queue situations. For an insight into the correct
mathematical model, which could be studied analytically to derive its
properties, a statistical study is essential. Otherwise, the analytical
study would be divorced from the practical situation, thus rendering it
less useful for the applied researcher.

(3) <u>Optimization problems of the system</u>. Looking at the variety of
queueing systems possible, it is only natural to assume that some models are
more appropriate in certain contexts than the rest. Furhter, as queue
situations can be controlled according to specifications, the knowledge of
the right model for the right situation becomes essential. The suitability of
a model is decided by a comparison among several models of the return in

benefit to the individual concerned. Naturally, some cost factors have
to be considered. Because of the external factors involved, the internal
characteristics have to be changed according to these. Thus some decision
rules should be spelled out.

A queueing system can be studied under two different assumptions.
One would aim at the short term behavior of the system, in which the results
turn out to be time dependent (transient behavior). Instead, one may assume
that the system has been in progress sufficiently long, so that it has
settled into exhibiting a stable behavior. This can happen only under certain
conditions (such as the traffic intensity less than one). The results obtained
in this study are independent of time and hence are much more simple in form.
Because of this simplicity, investigators in need of queueing results, have
tended to use steady state results, in most of the situations. When the system
has not settled into an equilibrium state, at best, this approach would give
some approximations to the actual behavior.

The importance of the study of queueing systems in finite time was noted
as early as 1934 by Pollaczek. The difficulty in such a study is that the
processes involved are not simple and more sophisticated mathematical procedures
are necessary. For instance, the birth and death process equations are simple
enough in the case of a simple queue with Poisson arrivals and exponential
service times. But for a time dependent solution the use of transforms is
necessary. This solution was given for the first time by Bailey (1956) and

Lederman & Reuter (1956). While Bailey used the method of generating functions for the differential equations, Lederman & Reuter used Spectral Theory in its solution. Laplace transforms have also been used for the same problem, and it has been realized that generating functions and/or Laplace transforms form a useful technique in the solution of such difference -differential equations. Several systems have been studied by this method and for a complete reference on other methods we refer to the bibliographies given in the books by Syski (1960), Saaty (1961), Takács (1962), Le Gall (1962) and Prabhu (1965).

Other methods which rely on the heavy use of transforms are the Takács equation method (1955) (here, balance of state equations are written for the waiting time process giving an intgro - differential equation, to be solved by the use of transforms), the supplementary variable techniques of Cox (1955), Keilson and Kooharian (1960) (here, the non-Markovian processes are rendered Markovian by describing the process with sufficient number of supplementary variables and operating on them. In addition to this, Keilson and Kooharian and also Keilson in his later investigations, use first passage problems connected with sections of the process studied in isolation) and the techniques of recurrence relations and renewal theory used in varied forms by Gaver (1959), Takács (1962a) and Bhat (1964, 1967b). (Here, recurrence relations for transitions in between successive regeneration points of the process are constructed and the entire analysis is built on these relations, often using renewal theory arguments].

The aim of the study of a real system should be the improvement of the design of the system. Unless we are able to invert some of the complicated transforms obtained in some of the papers referred to above, they cannot be used for practical purposes. In view of this, an investigation into the possibilities of getting results without using transforms seems to be worthwhile.

In this memorandum we study two classes of single server queueing systems (i) with Poisson arrivals and general service times (M/G/1) and (ii) general independent arrivals and exponential service times (GI/M/1), for its queue length behavior in continuous time. The time dependent and steady state behavior of the queue length process forms the major part of the study and the behavior of the waiting time process follows as a corollary.

Our method is applicable even when the customers arrive in groups in M/G/1 and get served in groups in the system GI/M/1. So as not to lose this generality we shall deal with these systems with group arrivals or group service. For the analysis of the original systems, one has only got to assume that the group size is always unity with probability one.

In denoting the queueing systems we use Kendall's (1953) notation which represents interarrival time distribution/service time distribution/number of servers, in that order.

A complete bibliography on the subject, classified into different systems appears at the end of the memorandum. However, for the sake of completeness, it may be mentioned here that the imbedded Markov chain analysis of these systems by combinatorial arguments is given in Prabhu and Bhat (1963) which occurs as sections 3.1 - 3.7 of Prabhu (1965).

1. The Queue M/G/1 with Group Arrivals

1.1 Description and Definitions

Consider a service facility with only one server. Customers arrive in groups of size $\{G_n\}$ (n = 1,2...) with a distribution

$$Pr\{G_n = j\} = b_j \quad (j = 1,2...). \tag{1.1}$$

These group arrivals occur in a Poisson process with parameter λ so that, if $A(t)$ is the number of arrivals during $(0,t]$ we have

$$Pr\{A(t) = j\} = \sum_{k=0}^{j} e^{-\lambda t} \frac{(\lambda t)^k}{k!} b_j^{(k)} \tag{1.2}$$

where $b_j^{(k)}$ is the k-fold convolution of $\{b_j\}$ with itself and $b_j^{(0)} = 0$ for $j > 0$, $= 1$ for $j = 0$. Also let

$$G(z) = \sum_{1}^{\infty} b_r z^r$$

with $0 < G'(1) < \infty$, $G'(1)$ being its mean.

Let the service times $\{v_n\}$ (n = 1,2...) of customers be distributed as $Pr\{v_n \leq x\} = B(x)$ with its Laplace-Stieltjes transform given by

$$\psi(\theta) = {}_0\!\int^{\infty} e^{-\theta t} dB(t) \quad (Re(\theta) \geq 0)$$

and $0 < -\psi'(0) < \infty$, where $-\psi'(0)$ is its mean. Consequently, the relative traffic intensity ρ is given by

$$\rho = -\lambda G'(1)\psi'(0).$$

The service mechanism is such that the server is idle only when there is no customer in the system. When there are customers waiting, the server takes them for service one after another in a manner determined by the queue discipline. A customer arriving at the system when the server is idle enters into the service immediately.

Let $Q(t)$ be the number of customers in the system at time t (we call this 'queue length') and for the sake of convenience we shall assume that $t = 0$ is an epoch of commencement of service. If $t = 0$ is not such a point, then it is possible to modify our results with the knowledge of the distribution of the unexpended service time of the customer at the counter. Further, we are concerned with the total number of customers present in the system, and therefore, the manner in which they enter service (such as, 'first come, first served', 'random selection for service' etc.) is immaterial as long as the arrival scheme and the service mechanism remain as specified above. Only when we consider the waiting time these distinctions are to be made.

Let $N(t)$ be the number of customers served in $(0,t]$ and define the random variable T_i as

$$T_i = \inf \ \{t \,|\, Q(t) = 0\} \ , \ Q(0) = i; \tag{1.3}$$

This represents the length of the busy period initiated by a queue length i.

1.2 Busy Period Transitions

Let

$$G_i^{(n)}(t) = Pr\{T_i \leq t , N(T_i) = n\}, \tag{1.4}$$

the joint distribution of the length of and the number of customers served in a busy period initiated by i customers. We have

Theorem 1.1

$$dG_i^{(n)}(t) = \frac{i}{n} e^{-\lambda t} \sum_{k=0}^{n-i} \frac{(\lambda t)^k}{k!} b_{n-i}^{(k)} dB_n(t) \quad (n \geq i , i > 0) \tag{1.5}$$

where $B_n(t)$ is the n-fold convolution of $B(t)$ with itself with $B_o(t) = 0$ for $t > 0$ and $= 1$ for $t = 0$.

Proof: We shall first prove the following identity which shall be used later in the proof of the theorem. We have

$$\int_0^t \tau \, dB_n(\tau) dB_m(t-\tau) = \frac{nt}{m+n} dB_{m+n}(t), \tag{1.6}$$

[Prabhu (1960 a), equation (6)].

For, the Laplace transform of the left hand side can be written as

$$\int_0^\infty e^{-\theta t} \int_0^t \tau \, dB_n(\tau) dB_m(t-\tau) = [\psi(\theta)]^m \int_0^\infty \tau e^{-\theta \tau} dB_n(\tau)$$

$$= -n[\psi(\theta)]^{m+n-1}\psi'(\theta)$$

$$= \frac{n}{m+n} \int_0^\infty t \, e^{-\theta t} dB_{m+n}(t) \tag{1.7}$$

which is the Laplace transform of the right hand side.

Now consider the probability $dG_i^{(n+i)}(t)$. Let $\tau(0<\tau\leq t)$ be the time at which the initial i customers complete their service. If the busy period extneds beyond this time, $r(=1,2,\ldots n)$ customers would have arried during the interval $(0,\tau]$ and therefore we have the recurrence relations

$$dG_i^{(i)}(t) = e^{-\lambda t}dB_i(t) \quad (i \geq 1) \tag{1.8}$$

$$dG_i^{(n+i)}(t) = \sum_{r=1}^{n} \sum_{k=1}^{r} \int_0^t e^{-\lambda\tau} \frac{(\lambda\tau)^k}{k!} \, b_r^{(k)} dB_i(t) dG_r^{(n)}(t-\tau) \tag{1.9}$$
$$(n > 0).$$

Setting $n = 1$ in (1.9) and using (1.8) we get

$$dG_i^{(i+1)}(t) = \int_0^t e^{-\lambda\tau} \, \lambda\tau b_1 dB_i(\tau) e^{-\lambda(t-\tau)} dB_1(t-\tau)$$
$$= \frac{i}{i+1} e^{-\lambda t} \lambda t \, b_1 \, dB_{i+1}(t), \tag{1.10}$$

where we have used the result (1.6) with $n = i$ and $m = 1$. Now we assert that the solution of (1.9) is given by

$$dG_i^{(n)}(t) = \frac{i}{n} e^{-\lambda t} \sum_{k=0}^{n-i} \frac{(\lambda t)^k}{k!} \, b_{n-i}^{(k)} \, dB_n(t). \tag{1.11}$$

Clearly, this is true for the sets of values (i,i) and $(i,i+1)$ of (i,n). Assuming it to be true for (r,n), $r = 1,2\ldots n$ and substituting in (1.9) we have

$$dG_i^{(i+n)}(t) = e^{-\lambda t} \int_0^t dB_i(\tau) dB_n(t-\tau)$$

$$\sum_{r=1}^{n} \frac{r}{n} \sum_{k=1}^{r} \sum_{s=0}^{n-r} \frac{(\lambda\tau)^k}{k!} \frac{[\lambda(t-\tau)]^s}{s!} \, b_r^{(k)} \, b_{n-r}^{(s)}$$

$$= e^{-\lambda t} \, {}_0\!\int^t dB_i(\tau)dB_n(t-\tau)$$

$$\frac{1}{n} \sum_{s=1}^{n} \sum_{k=0}^{n-s} \frac{(\lambda\tau)^s}{s!} \frac{[\lambda(t-\tau)]^k}{k!} \sum_{r=s}^{n-k} rb_r^{(s)} b_{n-r}^{(k)}$$

$$= e^{-\lambda t} \, {}_0\!\int^t dB_i(\tau)dB_n(t-\tau)$$

$$\sum_{s=1}^{n} \sum_{k=0}^{n-s} \frac{(\lambda\tau)^s}{s!} \frac{[\lambda(t-\tau)]^k}{k!} \frac{b_n^{(s+k)}}{s+k}$$

$$= e^{-\lambda t} \, {}_0\!\int^t \tau dB_i(\tau)dB_n(t-\tau) \sum_{k=1}^{n} \frac{\lambda^k t^{k-1}}{k!} b_n^{(k)}$$

$$= \frac{i}{n+i} e^{-\lambda t} \sum_{k=1}^{n} \frac{(\lambda t)^k}{k!} b_n^{(k)} dB_{n+i}(t) \qquad (1.12)$$

which proves our assertion. In our simplifications, we have used the identity (1.6) and its discrete analogue,

$$\sum_{r=0}^{n} r \, b_r^{(s)} \, b_{n-r}^{(k)} = \frac{sn}{s+k} \, b_n^{(s+k)}. \qquad (1.13)$$

The joint distribution of the length of and the number of customers served in a conventionally defined busy period is obtained by setting $i = 1$ in (1.5). In this case the result is due to Gaver (1959) [equation 3.16b]. (Also see equation (6), p. 110 of Prabhu (1965)). Another important result connected with the busy period is given by the following.

Theorem 1.2

$$Pr\ \{T_i < \infty\} = \begin{cases} 1 & \text{if } \rho \leq 1 \\[2em] \zeta^i & \text{if } \rho > 1 \end{cases} \qquad (1.14)$$

where ζ is the smallest positive real root of the equation

$$z = \psi(\lambda - \lambda G(z)) \qquad (1.15)$$

with $\quad \psi(\theta) = {}_0\!\int^\infty e^{-\theta x} dB(x) \qquad (\text{Re}(\theta) \geq 0)$

and $\quad G(z) = \sum_1^\infty b_r z^r.$

Proof: To prove (1.14) we shall first obtain the Laplace transform of the busy period distribution.

Let $\ E(e^{-\theta T_i}\omega^{N(T_i)}) = \Pi_i(\theta,\omega)\ ,\ (\text{Re}(\theta) \geq 0\ ,\ |\omega| \leq 1) \qquad (1.16)$

Representing the busy period as a vector random variable, we have from equation (1.9)

$$(T_i\ ,\ N(T_i)) = (V_i\ ,\ i) + (T_{A(V_i)}\ ,\ N(T_{A(V_i)})) \qquad (1.17)$$

where $\ V_i = v_1 + v_2 \ldots + v_i\ $ and $\ A(t)\ $ is the number of arrivals during the interval $(0,t]$. Thus we can write

$$\Pi_i(\theta,\omega) = \sum_{j=0}^\infty E(e^{-\theta V_i}\omega^i)E(e^{-\theta T_j}\omega^{N(T_j)}|A(V_i) = j) \qquad (1.18)$$

We further note that $\ A(t)\ $ and $\ V_i\ $ are processes with stationary independent increments; therefore we can write

$$(T_i, N(T_i)) \sim \sum_{k=1}^i (v_k, 1) + \sum_{k=1}^i (T_{A(\nu_k)}, N(T_{A(\nu_k)}))$$

$$\sim (T_1, N(T_1)) + (T_1, N(T_1)) + \ldots (T_1, N(T_1)) \qquad (1.19)$$

$$(\text{i terms})$$

where the sign $\sim$ denotes probabilistic equivalence. This leads to

$$\Pi_i(\theta,\omega) = [\Pi_1(\theta,\omega)]^i. \qquad (1.20)$$

Using this result in (1.18) we get

$$[\Pi_1(\theta,\omega)]^i = \sum_{j=0}^{\infty} \sum_{k=0}^{j} \int_0^{\infty} \omega^i e^{-\theta t} [\Pi_1(\theta,\omega)]^j e^{-\lambda t} \frac{(\lambda t)^k}{k!} b_j^{(k)} dB_i(t)$$

$$= \omega^i[\psi(\theta+\lambda-\lambda G(\Pi_1(\theta,\omega)))]^i \qquad (1.21)$$

which shows that $z = \Pi_1(\theta,\omega)$ must satisfy the functional equation

$$z = \omega\psi(\theta+\lambda-\lambda G(z)). \qquad (1.22)$$

Let $z = \gamma(\theta,\omega)$ be this root. In fact, it can be shown that for $\text{Re}(\theta) \geq 0$, $|\omega| < 1$ or $\text{Re}(\theta) > 0$, $|\omega| \leq 1$ or $\text{Re}(\theta) \geq 0$, $|\omega| \leq 1$ and $\rho > 1$, $\gamma(\theta,\omega)$ is the only root of (1.22) in the unit circle $|z| < 1$. For, when $\delta > 0$ and sufficiently small, when the conditions stated above are satisfied, $|\omega\psi(\theta+\lambda-\lambda G(z))| < |z|$ if $|z| = 1-\delta$, and the conclusion follows from Rouche's theorem. [For detailed discussion of the roots of equations of the type of (1.22), see Takács (1962), Lemma 1, page 47]. Thus we get

$$\Pi_i(\theta,\omega) = [\gamma(\theta,\omega)]^i \qquad (1.23)$$

Setting $\omega = 1$, we obtain the Laplace transform of the distribution of the length of the busy period, and therefore

$$\Pr\{T_i < \infty\} = \lim_{\theta \to 0+} [\gamma(\theta,1)]^i. \qquad (1.24)$$

Consider the equation (1.22) for $\omega = 1$ and $\theta = 0$. We have

$$z = \psi(\lambda-\lambda G(z)).$$

Instead, we consider, $f(z) = \psi(\lambda-\lambda G(z))-z.$ \qquad (1.25)

We have,

$$f'(z) = -\lambda G'(z)\psi'(\lambda-\lambda G(z)) - 1$$

$$f''(z) = -\lambda G''(z)\psi'(\lambda-\lambda G(z)) + [\lambda G'(z)]^2\psi''(\lambda-\lambda G(z))$$

$$= \lambda G''(z) \int_0^\infty x\, e^{-(\lambda-\lambda G(z))x}\, dB(x)$$

$$+ [\lambda G'(z)]^2 \int_0^\infty x^2\, e^{-(\lambda-\lambda G(z))x}\, dB(x) > 0 \qquad (1.26)$$

showing that $f'(z)$ is monotone increasing.

Further,

$$f(0) = \psi(\lambda) > 0$$

$$f(1) = 0$$

and
$$f'(1) = -\lambda G'(1)\psi'(0)-1 = \rho-1 \qquad (1.27)$$

Consequently, if $\rho>1$, there is only one real root $\zeta(0<\zeta<1)$ of the equation $f(z) = 0$ and if $\rho \leq 1$, $z = 1$ is such a root. Thus the root we are looking for is the smallest positive real root of the equation $f(z) = 0$. This completes the proof of the theorem.

For the complete understanding of the behavior of the stochastic process $Q(t)$ we need to study

$$^0P_{ij}^{(n)}(t) = \Pr\{Q(t)=j\ ;\ N(t)=n\ ,\ T_i>t\,|\,Q(0)=i\}\ ;$$

$$(i,\ j \geq 1), \qquad (1.28)$$

We shall call $^0P_{ij}^{(n)}(t)$ as the zero-avoiding transition probabilities. To obtain (1.28) we proceed as follows.

Consider the Poisson process $A(t)$ defined as in (1.2) and the renewal process $\{V_n\}$, $V_n = v_1 + v_2 + \ldots + v_n$, where the $v_n(n=1,2\ldots)$ have the distribution $dB(v)$. Let $D(t) = \max \{n|V_n \leq t\}$, so that

$$Pr\{D(t)=n\}=B_n(t) - B_{n+1}(t) = C_n(t) \text{ (say), } (n \geq 0). \qquad (1.29)$$

It is clear that the r.v. T_i representing the length of the busy period can be defined with reference to the processes $A(t)$ and $D(t)$ as

$$T_i = \inf\{t|i + A(t) - D(t) \leq 0\}. \qquad (1.30)$$

The process $i + A(t) - D(t)$ is non-Markovian; however, the points at which $T_i = \tau$ (for some τ) are points of regeneration. We have the following

Lemma 1.1:

$$Pr\{i + A(t) - D(t) = j , T_i > T ; D(t) = n\}$$

$$= e^{-\lambda t} \sum_{k=0}^{n+j-i} \frac{(\lambda t)^k}{k!} b_{n+j-i}^{(k)} C_n(t)$$

$$- \sum_{m=i}^{n} \int_0^t Pr\{\tau<T_i < \tau + d\tau ; D(\tau) = m\}$$

$$e^{-\lambda(t-\tau)} \sum_{k=1}^{n-m+j} \frac{[\lambda(t-\tau)]^k}{k!} b_{n-m+j}^{(k)} C_{n-m}(t-\tau)$$

$$\qquad (1.31)$$

Proof: We have

$$e^{-\lambda t} \sum_{k=0}^{n+j-i} \frac{(\lambda t)^k}{k!} b_{n+j-i}^{(k)} C_n(t) = \Pr\{i+A(t)-D(t)=j \ ; \ D(t)=n\}$$

$$= \Pr\{i+A(t)-D(t)=j \ , \ T_i > t \ ; \ D(t)=n\}$$

$$+ \Pr\{i+A(t)-D(t)=j \ , \ T_i \le t \ ; \ D(t)=n\}. \tag{1.32}$$

The first term on the right hand side of (1.32) is the required probability, while the second term can be written as

$$= \sum_{m=i}^{n} \int_0^t \Pr\{\tau < T_i < \tau + d\tau \ ; \ D(\tau)=m\}.$$

$$\Pr\{i+A(t)-D(t)=j \ ; \ D(t)=n | T_i=\tau \ , \ D(\tau)=m\}$$

$$= \sum_{m=i}^{n} \int_0^t \Pr\{\tau < T_i < \tau + d\tau \ ; \ D(\tau)=m\}$$

$$\Pr\{i+A(t)-D(t)=j \ ; \ D(t)=n | i+A(\tau)-D(\tau) = 0 \ , \ D(\tau) = m\}$$

$$= \sum_{m=i}^{n} \int_0^t \Pr\{\tau < T_i < \tau + d\tau \ ; \ D(\tau)=m\} \ \Pr\{A(t-\tau)-D(t-\tau)=j \ ;$$

$$D(t-\tau)=n-m\}$$

$$= \sum_{m=i}^{n} \int_0^t \Pr\{\tau < T_i < \tau + d\tau \ ; \ D(\tau)=m\} \ e^{-\lambda(t-\tau)} \sum_{k=1}^{n-m+j}$$

$$\frac{[\lambda(t-\tau)]^k}{k!} b_{n-m+j}^{(k)}$$

$$C_{n-m}(t-\tau). \tag{1.33}$$

Using (1.33) in (1.32) we have the Lemma. Incidentally we can also establish the following identity concerning the process $i+A(t)-D(t)$.

An identity: When $j \leq 0$, we have

$$e^{-\lambda t} \sum_{k=0}^{n+j-i} \frac{(\lambda t)^k}{k!} b_{n+j-i}^{(k)} C_n(t)$$

$$= \sum_{m=i}^{n+j} {}_0\!\int^t \Pr\{\tau < T_i < \tau + d\tau \; ; \; D(\tau) = m\} e^{-\lambda(t-\tau)} \sum_{k=0}^{n-m+j} \frac{[\lambda(t-\tau)]^k}{k!} b_{n-m+j}^{(k)}$$

$$C_{n-m}(t-\tau) \quad (n \geq i-j). \qquad (1.34)$$

This is obtained from the relation

$$\Pr\{i+A(t)-D(t)=j \; ; \; D(t)=n\} = e^{-\lambda t} \sum_{k=0}^{n+j-i} \frac{(\lambda t)^k}{k!} b_{n+j-i}^{(k)} C_n(t)$$

which is also

$$= \sum_{m=i}^{n+j} {}_0\!\int^t \Pr\{\tau < T_i < \tau + d\tau \; ; \; D(\tau) = m\}.$$

$$\Pr\{i+A(t)-D(t)=j \; ; \; D(t)=n \,|\, T_i = \tau \; ; \; D(\tau) = m\} \; , \; (j \leq 0) \qquad (1.35)$$

This identity can be used in obtaining the Laplace transform of the distribution ${}^0P_{ij}^{(n)}(t)$, which is given by

Theorem 1.3

$${}^0P_{ij}^{(n)}(t) = e^{-\lambda t} \sum_{k=0}^{n+j-i} \frac{(\lambda t)^k}{k!} b_{n+j-i}^{(k)} C_n(t)$$

$$- \sum_{m=0}^{n-i} {}_0\!\int^t dG_i^{(n-m)}(\tau) \, e^{-\lambda(t-\tau)}$$

$$\sum_{k=0}^{m+j} \frac{[\lambda(t-\tau)]^k}{k!} b_{m+j}^{(k)} C_m(t-\tau)$$

[Prabhu and Bhat (1963)]

<u>Proof</u>: We note that if $T_i > t$, the queue length $Q(t) = i + A(t) - D(t)$. Using the Lemma derived above, we get

$$
{}^0P_{ij}^{(n)}(t) = \Pr\{i + A(t) - D(t) = j \ , \ D(t) = n \ ; \ T_i > t\}
$$

$$
= e^{-\lambda t} \sum_{k=0}^{n+j-i} \frac{(\lambda t)^k}{k!} \, b_{n+j-i}^{(k)} \, C_n(t)
$$

$$
- \sum_{m=i}^{n} \int_0^t \Pr\{\tau < T_i < \tau + d\tau \ ; \ D(\tau) = m\} e^{-\lambda(t-\tau)}
$$

$$
\sum_{k=1}^{n-m+j} \frac{[\lambda(t-\tau)]^k}{k!} \cdot b_{n-m+j}^{(k)} \, C_{n-m}(t-\tau). \tag{1.37}
$$

The second term in (1.37) can be simplified as

$$
\sum_{m=0}^{n-i} \int_0^t \Pr\{\tau < T_i < \tau + d\tau \ ; \ D(\tau) = n-m\} e^{-\lambda(t-\tau)} \sum_{k=1}^{m+j} \frac{[\lambda(t-\tau)]^k}{k!} \, b_{m+j}^{(k)} \, C_m(t-\tau)
$$

$$
= \sum_{m=0}^{n-i} \int_0^t dG_i^{(n-m)}(\tau) e^{-\lambda(t-\tau)} \sum_{k=1}^{m+j} \frac{[\lambda(t-\tau)]^k}{k!} \, b_{m+j}^{(k)} \, C_m(t-\tau). \tag{1.38}
$$

This proves the theorem.

In the next section, we shall use the results of Theorems 1.1 and 1.3 after summing over n. We shall denote these probabilities by

$$
dG_i(t) = \sum_{n=i}^{\infty} dG_i^{(n)}(t) \tag{1.39}
$$

and

$$
{}^0P_{ij}(t) = \sum_{n=0}^{\infty} {}^0P_{ij}^{(n)}(t). \tag{1.40}
$$

Accordingly we shall set $\omega = 1$ in the functional equation (1.22) and write $\gamma(\theta)$ instead of $\gamma(\theta, 1)$.

1.3 General Transitions of Q(t)

Let $P_{ij}(t)$ denote the general transition probability of the queue length $Q(t)$ defined by

$$P_{ij}(t) = Pr\{Q(t)=j \mid Q(0)=i\} \qquad (i, j \geq 0). \qquad (1.41)$$

We shall first obtain the probability $P_{io}(t)$ $(i \geq 0)$ using renewal theroy and establish certain relations to give $P_{ij}(t)$ $(i \geq 0, j > 0)$ in terms of known probabilities.

Consider a sequence of independent r.v's $T_i^{(0)}$, I_1, $T^{(1)}$, I_2, $T^{(2)}$,...; let $I_r(r=1, 2...)$ be distributed identically as

$$dI(t) = Pr\{t < I_r < t+dt\} = \lambda e^{-\lambda t} dt \quad (0 < t < \infty), \qquad (1.42)$$

$T_i^{(0)}$ as $dG_i(t)$ and $T^{(r)}$ $(r=1, 2...)$ as $\sum_r b_r dG_r(t)$. Referring to the queueing process, I_1, I_2... refer to the consecutive idle periods and $T_i^{(0)}$, $T^{(1)}$, $T^{(2)}$,... to the consecutive busy periods $(T_i^{(0)}$ is the initial busy period); in this process idle periods and busy periods follow one after the other alternately. It should also be noted that in the system M/G/1, I_r is independnet of $T^{(r-1)}$ because of the reproducing property of the conditional density of the negative exponential distribution. Define a sequence of r.v.'s $\{X_n\}$ $(n=0, 1...)$ such that $X_o = T_i^{(0)}$ and $X_n = I_n + T^{(n)}$ $(n=1,2...)$. Clearly $\{X_n\}$ is a renewal process. Let $S_n = X_1 + X_2 + ... + X_n$, and $S_n^{(i)} = X_0 + X_1 + ... + X_n$; we have

$$S_n = I_1 + I_2 + ... + I_n + T^{(1)} + T^{(2)} + ... + T^{(n)}$$

$$\wr T_n + I^{(n)}$$

and
$$S_n^{(i)} \wr T_{n+i} + I^{(n)} \qquad\qquad (1.43)$$

where $I^{(n)}$ has the distribution

$$dI^{(n)}(t) = e^{-\lambda t} \frac{(\lambda t)^{n-1}}{(n-1)!} \lambda dt \qquad (0 < t < \infty). \qquad (1.44)$$

Define the renewal functions

$$U(t) = \sum_{n=1}^{\infty} \Pr\{S_n \leq t\}$$

$$U_i(t) = \sum_{n=0}^{\infty} \Pr\{S_n^{(i)} \leq t\}. \qquad (1.45)$$

For these we have the following

Lemma 1.2:

a) $\quad dU(t) = e^{-\lambda t} \sum_{n=1}^{\infty} \sum_{k=1}^{n} \frac{\lambda^k t^{k-2}}{(k-1)!} b_n^{(k)} \int_0^t [t - (1-\frac{1}{k})\tau] dB_n(\tau) dt$

$$\qquad\qquad\qquad\qquad\qquad\qquad\qquad\qquad\qquad (1.46)$$

b) $\quad dU_i(t) = e^{-\lambda t} \sum_{n=i}^{\infty} \sum_{k=0}^{n-i} \frac{i}{n} \frac{(\lambda t)^k}{k!} b_{n-i}^{(k)} dB_n(t)$

$$\qquad\qquad + e^{-\lambda t} \sum_{n=i+1}^{\infty} \sum_{k=1}^{n-i} \frac{\lambda^k t^{k-2}}{(k-1)!} b_{n-i}^{(k)} \int_0^t$$

$$\qquad\qquad\qquad\qquad [t - \frac{(n-i)(k-1)}{nk} \tau] dB_n(\tau) dt \quad (1.47)$$

Proof: Using the expressions given by (1.43) for S_n and $S_n^{(i)}$ we can write

$$dU(t) = \sum_{s=1}^{\infty} \sum_{n=1}^{s} \int_0^t dG_s(\tau) dI^{(n)}(t-\tau) b_s^{(n)} \qquad (1.48)$$

and

$$dU_i(t) = dG_i(t) + \sum_{s=1}^{\infty} \sum_{n=1}^{s} {}_0\!\int^t dG_{i+s}(\tau) dI^{(n)}(t-\tau) b_s^{(n)} \qquad (1.49)$$

where $dG_i(t)$ and $dI^{(n)}(t)$ are given by (1.39) and (1.44) respectively. It should be noted that the first term on the right hand side of (1.49) accounts for the probability of $n=0$ in the definition (1.45) of $U_i(t)$. Since (1.48) is a special case of (1.49) with $i=0$, it is enough to deal only with the latter.

The second term in (1.49) can be written as

$$= \sum_{s=1}^{\infty} \sum_{n=1}^{s} {}_0\!\int^t \sum_{r=i+s}^{\infty} \sum_{k=0}^{r-i-s} \frac{i+s}{r} e^{-\lambda\tau} \frac{(\lambda\tau)^k}{k!} b_{r-i-s}^{(k)} dB_r(\tau)$$

$$e^{-\tau(t-\tau)} \frac{[\lambda(t-\tau)]^{n-1}}{(n-1)!} \lambda dt\, b_s^{(n)}$$

After changing summations we get this

$$= \lambda e^{-\lambda t} dt\, {}_0\!\int^t \sum_{n=1}^{\infty} \frac{[\lambda(t-\tau)]^{n-1}}{(n-1)!} \sum_{r=i+n}^{\infty} \frac{1}{r} dB_r(\tau)$$

$$\sum_{k=0}^{r-i-n} \frac{(\lambda\tau)^k}{k!} \sum_{s=n}^{r-i-k} (i+s)\, b_{r-i-s}^{(k)}\, b_s^{(n)}$$

$$= \lambda e^{-\lambda t} dt\, {}_0\!\int^t \sum_{r=i+1}^{\infty} \frac{1}{r} dB_r(\tau) \sum_{n=1}^{r-i} \frac{[\lambda(t-\tau)]^{n-1}}{(n-1)!} \sum_{k=n}^{r-i} \frac{(\lambda\tau)^{k-n}}{(k-n)!}$$

$$b_{r-i}^{(k)} [i + \frac{(r-i)n}{k}]$$

$$= \lambda e^{-\lambda t} dt \; {}_0\!\int^t \sum_{r=i+1}^{\infty} \frac{1}{r} \, dB_r(\tau) \sum_{k=1}^{r-i} \lambda^k b_{r-i}^{(k)}$$

$$\sum_{n=1}^{k} [i + \frac{(r-i)n}{k}] \; \frac{\tau^{k-n}(t-\tau)^{n-1}}{(k-n)!(n-1)!}$$

which after some simplifications gives

$$\lambda e^{-\lambda t} dt \; {}_0\!\int^t \sum_{r=i+1}^{\infty} \frac{1}{r} \, dB_r(\tau) \sum_{k=1}^{r-i} \lambda^k b_{r-i}^{(k)} \; [\; \frac{rt^{k-1}}{(k-1)!} - \frac{(r-i)(k-1)t^{k-2}}{k!} \; \tau]$$

which can be seen to be the same as the second term on the right hand

side of (1.49), after some rearrangement.

We are now in a position to give

<u>Theorem 1.4</u>: For $i \geq 0$,

$$P_{io}(t) = \sum_{n=i}^{\infty} \sum_{k=0}^{n-i} e^{-\lambda t} \frac{\lambda^k}{k!} b_{n-i}^{(k)} \, t^{k-1} \; {}_0\!\int^t [t - (1-\frac{i}{n})\tau] dB_n(\tau)$$

$$(1.51)$$

<u>Proof</u>:

Case (i): i=0

The transition $0 \to 0$ in time $(0,t]$ can occur, either

with no arrival or with at least one arrival in $(0,t]$. In the latter

case let $\tau(0 < \tau < t)$ be the time at which $Q(t)$ becomes zero

for the last time. These alternatives give

$$P_{oo}(t) = e^{-\lambda t} + {}_0\!\int^t dU(\tau)e^{-\lambda(t-\tau)}$$

$$(1.52)$$

where $dU(t)$ is given by (1.46).

Case (ii): i > 0

Consider the time $\tau(0 < \tau < t)$ at which $Q(t)$ becomes zero for the last time. Considering the transitons before and after τ , we have

$$P_{io}(t) = {}_0\!\int^t dU_i(\tau)e^{-\lambda(t-\tau)} \tag{1.53}$$

where $dU_i(t)$ is given by (1.47).

As the methods of simplifications of (1.52) and (1.53) are the same, leading to similar results, we shall simplify only the latter. We have,

$$P_{io}(t) = \sum_{n=i}^{\infty} \sum_{k=0}^{n-i} \frac{i}{n} e^{-\lambda t} \frac{\lambda^k}{k!} b_{n-i}^{(k)} {}_0\!\int^t \tau^k dB_n(\tau)$$

$$+ \sum_{n=i+1}^{\infty} \sum_{k=1}^{n-i} e^{-\lambda t} \frac{\lambda^k}{(k-1)!} b_{n-i}^{(k)} {}_0\!\int^t \tau^{k-2} d\tau {}_0\!\int^\tau$$

$$[\tau - \frac{(n-i)(k-1)}{nk} s]\ dB_n(s) \tag{1.54}$$

The second term in (1.54) simplifies to

$$\sum_{n=i+1}^{\infty} \sum_{k=1}^{n-i} e^{-\lambda t} \frac{\lambda^k}{(k-1)!} b_{n-i}^{(k)} {}_0\!\int^t dB_n(s) {}_{\tau=s}\!\int^t \tau^{k-2}$$

$$[\tau - \frac{(n-i)(k-1)}{nk} s]d\tau$$

$$= \sum_{n=i+1}^{\infty} \sum_{k=1}^{n-i} e^{-\lambda t} \frac{\lambda^k}{k!} b_{n-i}^{(k)} t^{k-1} {}_0\!\int^t [t - (1-\frac{i}{n})\tau]\ dB_n(\tau)$$

$$- \sum_{n=i+1}^{\infty} \sum_{k=1}^{n-i} e^{-\lambda t} \frac{\lambda^k}{n!} b_{n-i}^{(k)} \frac{i}{nk} {}_0\!\int^t \tau^k dB_n(\tau). \tag{1.55}$$

The second term in (1.55) cancels with the first term in (1.54) leaving $e^{-\lambda t} \, {}_0\!\int^t dB_i(\tau)$ which when merged with the first term in (1.55) gives the required result.

Finally, the relations that give the general transition probabilities $P_{ij}(t)$ $(j \geq 1)$, are given by [Prabhu and Bhat (1963)].

Theorem 1.5 For $j \geq 1$.

a) $P_{ij}(t) = {}^oP_{ij}(t) + \lambda \, {}_0\!\int^t P_{io}(\tau) \sum_{r=1}^{\infty} b_r \, {}^oP_{rj}(t-\tau) d\tau \quad (i \geq 1)$ \hfill (1.56)

b) $P_{oj}(t) = \lambda \, {}_0\!\int^t P_{oo}(\tau) \sum_{r=1}^{\infty} b_r \, {}^oP_{rj}(t-\tau) d\tau.$ \hfill (1.57)

Proof: If $Q(t) = j \geq 1$, it is clear that a busy period is in progress at time t. In the case $i \geq 1$, this might be the initial one itself, or the one that commenced at time $\tau (0 < \tau < t)$ with a group of arrivals; these two probabilities are accounted for on the right hand side of (1.56). If $i = 0$ the first possibility does not arise.

1.4 Limiting Behavior of $Q(t)$

In this section we shall obtain

$$P_j^* = \lim_{t \to \infty} P_{ij}(t) \hfill (1.58)$$

Theorem 1.6:

$\lim_{t \to \infty} P_{ij}(t)$ is independent of the initial state i, and is given by

$$P_o^* = \begin{cases} 0 & \text{if } \rho \geq 1 \\[2ex] 1-\rho & \text{if } \rho < 1 \end{cases} \hfill (1.59)$$

and for $j > 0$,

$$P_j^* = \begin{cases} 0 & \text{if } \rho \geq 1 \\ \\ (1-\rho) \int_0^\infty e^{-\lambda t} \sum_{n=0}^\infty dB_n(t) \left[\sum_{k=1}^{n+j-1} b_{n+j}^{(k+1)} \frac{(\lambda t)^k}{k!} \right. \\ \\ \left. - \sum_{k=1}^{n+j} b_{n+j}^{(k)} \frac{(\lambda t)^k}{k!} \right] & \text{if } \rho < 1. \end{cases} \qquad (1.60)$$

Proof: We shall first obtain $\lim_{t \to \infty} P_{oo}(t)$ from its Laplace transform. Consider equation (1.48) for $dU(t)$. We have

$$\int_0^\infty e^{-\theta t} dG_i(t) = [\gamma(\theta)]^i$$

and

$$\int_0^\infty e^{-\theta t} dI^{(n)}(t) = \left(\frac{\lambda}{\lambda \theta + \lambda} \right)^n$$

where $\gamma(\theta)$ satisfies the functional equation $z = \psi(\theta + \lambda - \lambda G(z))$. Consequently

$$\int_0^\infty e^{-\theta t} dU(t) = \sum_{s=1}^\infty \sum_{n=1}^s b_s^{(n)} [\gamma(\theta)]^s \left(\frac{\lambda}{\theta + \lambda} \right)^n$$

$$= \sum_{n=1}^\infty \left[\frac{\lambda G(\gamma(\theta))}{\theta + \lambda} \right]^n$$

$$= \frac{\lambda G(\gamma(\theta))}{\theta + \lambda - \lambda G(\lambda(\theta))} \qquad (1.61)$$

where $G(z) = \sum_{r=1}^\infty b_r z^r$.

Now consider equation (1.52) for $P_{oo}(t)$. We have

$$\int_0^\infty e^{-\theta t} P_{oo}(t) dt = \frac{1}{\theta+\lambda} [1 + \frac{\lambda G(\gamma(\theta))}{\theta+\lambda-\lambda G(\gamma(\theta))}]$$

$$= \frac{1}{\theta+\lambda-\lambda G(\gamma(\theta))} \qquad (1.62)$$

From an Abelian Theorem (Widder, Chapter V).

$$\lim_{t\to\infty} P_{oo}(t) = \lim_{\theta\to 0+} \theta \int_0^\infty e^{-\theta t} P_{oo}(t) dt$$

$$= \lim_{\theta\to 0+} \frac{\theta}{\theta+\lambda-\lambda G(\gamma(\theta))} \qquad (1.63)$$

$$= 0 \qquad \text{for} \qquad \rho > 1$$

Since $\gamma(0+) = \xi < 1$ in this case. [See discussion under Theorem 1.2].

For $\rho \le 1$, $\gamma(0+) = 1$ and hence differentiating numerator as well as the denominator of (1.63)

$$\lim_{t\to\infty} P_{oo}(t) = \lim_{\theta\to 0+} \frac{1}{1-\lambda G'(1)\gamma'(0+)} \qquad (1.64)$$

To obtain $\gamma'(0+)$ we differentiate $\gamma(\theta) = \psi(\theta+\lambda-\lambda G(\gamma(\theta)))$ and make $\theta\to 0+$, with $\rho \le 1$. We get

$$\gamma'(0+) = - \frac{\rho}{\lambda G'(1)[1-\rho]} \qquad (1.65)$$

Combining (1.64) and (1.65) we have

$$\lim_{t\to\infty} P_{oo}(t) = 1 - \rho. \qquad (1.66)$$

It can be easily seen that this is also the limit with an initial
queue length $Q(t) > 0$. This can be established either by obtaining
the Laplace transform of $P_{io}^{\bullet}(t)$ as $[\gamma(\theta)]^i[\theta+\lambda-\lambda G(\gamma(\theta))]^{-1}$ and
obtaining the limit as before or using (1.66) in the obvious relation

$$P_{io}(t) = {}_0\!\int^t dG_i(\tau)P_{oo}(t-\tau) \qquad (1.67)$$

Here the limit can be obtained with an argument similar to the one
given below for the general transtion probability $P_{ij}(t)$. This
proves the first part of the theorem.

For the second part consider the general relation (1.56).
We have

$$\lim_{t\to\infty} P_{ij}(t) = \lim_{t\to\infty} {}^oP_{ij}(t) + \lim_{t\to\infty} \lambda {}_0\!\int^t P_{io}(\tau) \sum_{r=1}^{\infty} b_r \, {}^oP_{rj}(t-\tau)d\tau$$

$$(1.68)$$

Recalling that ${}^oP_{ij}(t)$ is given by (1.36) and (1.40) we can write

$${}^oP_{ij}(t) \le \sum_n e^{-\lambda t} \frac{(\lambda t)^{n+j-i}}{(n+j-i)!} \, C_n(t)$$

Now

$$e^{-\lambda t} \frac{(\lambda t)^{n+j-i}}{(n+j-i)!} \, C_n(t) \le e^{-\lambda t} \frac{(\lambda t)^{n+j-i}}{(n+j-i)!} \; ;$$

As $t \to \infty$, the Poisson probabilities in the last expression $\to 0$
and hence we have

$$\lim_{t\to\infty} {}^oP_{ij}(t) = 0.$$

We can simplify the second expression in (1.57) as follows.

Consider

$$\lambda \int_0^t P_{io}(\tau)d\tau \sum_{r=1}^{\infty} b_r \, {}^o P_{rj}(t-\tau) = \lambda \int_0^t P_{io}(t-\tau) \sum_{r=1}^{\infty} b_r \, {}^o P_{rj}(\tau)d\tau$$

$$= \lambda_0 \int_0^T P_{io}(t-\tau) \sum_{r=1}^{\infty} b_r \, {}^o P_{rj}(\tau)d\tau$$

$$+ \lambda_T \int_T^t P_{io}(t-\tau) \sum_{r=1}^{\infty} b_r \, {}^o P_{rj}(\tau)d\tau$$

$$= I_1 + I_2 \quad \text{(say)}. \tag{1.70}$$

In these, first let t and then $T \to \infty$; then

$$I_1 \to \lambda \lim_{t\to\infty} P_{io}(t) \int_0^{\infty} \sum_{r=1}^{\infty} b_r \, {}^o P_{rj}(\tau)d\tau$$

$$= \begin{cases} \lambda(1-\rho) \int_0^{\infty} \sum_{r=1}^{\infty} b_r \, {}^o P_{rj}(\tau)d\tau & \text{if } \rho < 1 \\ \\ 0 & \text{if } \rho \geq 1. \end{cases} \tag{1.71}$$

and

$$I_2 \leq \lambda_T \int_T^t \sum_{r=1}^{\infty} b_r \, {}^o P_{rj}(\tau)d\tau \to 0 \tag{1.72}$$

since the integrand is convergent over $(0,\infty)$. Further,

$$\int_0^{\infty} \sum_{r=1}^{\infty} b_r \, {}^o P_{rj}(t)dt = \int_0^{\infty} e^{-\lambda t} \sum_{n=0}^{\infty} \sum_{k=0}^{n+j-1} b_{n+j}^{(k+1)} \frac{(\lambda t)^k}{k!} C_n(t)dt$$

$$- \int_0^{\infty} \sum_{n=1}^{\infty} \sum_{m=0}^{n-1} \int_0^t \sum_{r=1}^{\infty} b_r \, dG_r^{(n-m)}(\tau)$$

$$e^{-\lambda(t-\tau)} \sum_{k=0}^{m+j} \frac{[\lambda(t-\tau)]^k}{k!} b_{m+j}^{(k)} C_m(t-\tau)dt. \tag{1.73}$$

The second expression on the right hand side of (1.73) can be further simplified as

$$_0\!\int^\infty \sum_{r=1}^\infty b_r dG_r(t) \; _0\!\int^\infty e^{-\lambda t} \sum_{n=0}^\infty \sum_{k=0}^{n+j} \frac{(\lambda t)^k}{k!} b_{n+j}^{(k)} C_n(t) dt$$

But we have

$$_0\!\int^\infty \sum_{r=1}^\infty b_r dG_r(t) = \sum_{r=1}^\infty b_r = 1 \tag{1.74}$$

Thus we get, for $j > 0$,

$$P_j^* = \lambda(1-\rho) \Big[\; _0\!\int^\infty e^{-\lambda t} \sum_{n=0}^\infty \sum_{k=0}^{n+j-1} b_{n+j}^{(k+1)} \frac{(\lambda t)^k}{k!} C_n(t) dt$$

$$- \; _0\!\int^\infty e^{-\lambda t} \sum_{n=0}^\infty \sum_{k=0}^{n+j} b_{n+j}^{(k)} \frac{(\lambda t)^k}{k!} C_n(t) dt \Big] \quad (\rho < 1) \tag{1.75}$$

This expression could be simplified to the form given by (1.60) by noting that

$$_0\!\int^\infty e^{-\lambda t} \sum_{k=0}^{n+j-1} b_{n+j}^{(k+1)} \frac{(\lambda t)^k}{k!} [B_n(t) - B_{n+1}(t)] dt$$

$$= \; _{\tau=0}\!\int^\infty dB_n(t) \; _{t=\tau}\!\int^\infty e^{-\lambda t} \sum_{k=0}^{n+j-1} b_{n+j}^{(k+1)} \frac{(\lambda t)^k}{k!} dt$$

$$- \; _{\tau=0}\!\int^\infty dB_{n+1}(t) \; _{t=\tau}\!\int^\infty e^{-\lambda t} \sum_{k=0}^{n+j-1} b_{n+j}^{(k+1)} \frac{(\lambda t)^k}{k!} dt \tag{1.76}$$

and

$$\lambda \; _{t=\tau}\!\int^\infty e^{-\lambda t} \sum_{k=0}^{n+j-1} b_{n+j}^{(k+1)} \frac{(\lambda t)^k}{k!} dt$$

$$= \sum_{k=0}^{n+j-1} b_{n+j}^{(k+1)} \sum_0^k e^{-\lambda \tau} \frac{(\lambda \tau)^r}{r!} \tag{1.77}$$

The relation (1.77) is obtained by repeated integration by parts of the left hand side.

The proof of the theorem is now complete.

1.5 Q(t) and the unexpended service time

In the study of the queue behavior at any time t, when the processes

involved are Markovian only at some points, as in the system M/G/1, the

intervals in which the processes are non-Markovian introduce further

difficulties some of which we have already seen. In particular, if one is

interested in the queue behavior at a time beyond t, it is essential,

that one also has enough information about the state of service at time t.

We shall call the time period, the customer at the counter will continue to

be served, as the unexpended service time.

Let S(t) be the unexpended service time of the customer at the counter,

at time t. The time t may be such that (i) there has been a departure

at τ $(0 \leq \tau < t)$ and $Q(\tau) = 0$ and (ii) there has been a departure at

τ $(0 \leq \tau < t)$ and $Q(\tau) > 0$. The possibility of no departure in (0,t) is

included in (i) and (ii) by setting $\tau = 0$.

Let $V(y,t) = Pr\{S(t) \leq y\}$. We have

$$V(y,t \mid \text{last departure at } \tau, Q(\tau) = 0) = \int_{s=\tau}^{t} \frac{B(t-s+y)-B(t-s)}{1-B(t-s)} \lambda e^{-\lambda(s-\tau)} ds$$

$$(1.78)$$

$$V(y,t \mid \text{last departure at } \tau, Q(\tau) > 0) = \frac{B(t-\tau+y)-B(t-\tau)}{1-B(t-\tau)} \ .$$

We can also write

$$V(y,t) = \int_{\tau=0}^{t} \left[\int_{s=\tau}^{t} \frac{B(t-s+y)-B(t-s)}{1-B(t-s)} \lambda e^{-\lambda(s-\tau)} ds \right.$$

$$+ \frac{B(t-\tau+y)-B(t-\tau)}{1-B(t-\tau)} \left. \right] d\tau \tag{1.79}$$

To simplify the notation, define $t*$ as a departure epoch (any one of the t_n's $n = 0,1,2,...$) and

$$^{o}P^{*}_{ij}(t) = \Pr\{Q(t*) = j, Q(\tau) > 0, 0 < \tau < t*, t* \le t | Q(0) = i\}$$
(1.80)

$$^{*}P_{ij}(t) = \Pr\{Q(t*) = j, t* \le t | Q(0) = i\}$$
(1.81)

and

$$P_{ij}(t,y) = \Pr\{Q(t) = j, S(t) \le y | Q(0) = i\}$$
(1.82)

Although we have not obtained $^{o}P^{*}_{ij}(t)$ and $^{*}P_{ij}(t)$ explicitly, expressions for these probabilities follow easily from what we have seen so far. Thus we can give

$$^{o}P^{*}_{ij}(t) = e^{-\lambda t} \sum_{n=0}^{\infty} \sum_{k=0}^{n+j-i} \frac{(\lambda t)^{k}}{k!} b^{(k)}_{n+j-i} B_{n}(t)$$

$$- \sum_{n=i}^{\infty} \sum_{m=0}^{n-i} {}_{0}\!\int^{t} dG_{i}^{(n-m)}(\tau) e^{-\lambda(t-\tau)}$$

$$\sum_{k=0}^{m+j} \frac{[\lambda(t-\tau)]^{k}}{k!} b^{(k)}_{m+j} B_{m}(t-\tau)$$
(1.83)

and

$$^{*}P_{ij}(t) = {}^{o}P^{*}_{ij}(t) + \lambda_{0}\!\int^{t} P_{io}(\tau) \sum_{r=1}^{\infty} b_{r} {}^{o}P^{*}_{rj}(t-\tau) d\tau \qquad (i \ge 1)(1.84)$$

Expressions for $^{*}P_{oj}(t)$ and $^{*}P_{io}(t)$ follow by similar arguments.

Now we have

Theorem 1.7: For $i \ge 0$, $j > 0$,

$$P_{ij}(t,y) = {}_{\tau=0^{-}}\!\int^{t} \sum_{r=1}^{j} d^{*}P_{ir}(\tau) e^{-\lambda(t-\tau)} \sum_{k=0}^{j-r} \frac{[\lambda(t-\tau)]^{k}}{k!} b^{(k)}_{j-r} \left[\frac{B(t-\tau+y)-B(t-\tau)}{1-B(t-\tau)} \right]$$

$$+ {}_{\tau=0^{-}}\!\int^{t} {}_{s=\tau}\!\int^{t} d^{*}P_{io}(\tau) \lambda e^{-\lambda(s-\tau)} ds e^{-\lambda(t-\tau)}$$

$$\sum_{k=0}^{j} \frac{[\lambda(t-\tau)]^{k}}{k!} b^{(k+1)}_{j+1} \left[\frac{B(t-s+y)-B(t-s)}{1-B(t-s)} \right]$$
(1.85)

Proof: Consider the last departure point τ in $[0,t)$ and transitions before and after this departure. Before this departure transitions are of the type $P^*_{ir}(\tau)$ and after the departure the transitions are of the type $V(y,t \mid \tau$ is a departure point). Now using expressions (1.84) and the two expressions in (1.78) we have the theorem. To allow for the possibility of no departure in $[0,t)$ we have included $\tau = 0$ as a possible value in (1.85).

In this and the following theorem, we shall not be simplifying the expressions derived by probabilistic arguments. As they are, it would be easier to simplify in special cases. If sufficiently sophisticated programs are available, these expressions are also suitable for comuter programming, as evidenced by the investigations carried out by the author on the transition probabilities given in Theorem 1.1 - 1.6.

1.6 Waiting time W(t): An approach through Q(t).

We define the waiting time process $W(t)$ as follows.

Suppose there is a mechanism (say, an inspector) that observes the system as to how long a new customer at time t, if any, will have to wait till he enters his service. Clearly, at any time t, with queue discipline 'first come-forst served' this would be the time required for all the customers in the system to complete their service. This is known as the virtual waiting time of the queueing system. Since $W(t)$ depends on the actual position of the customer in the queue, queue discipline is a deciding factor. We shall derive probabilistic relations between the processes $Q(t)$ and $W(t)$ in queueing systems with disciplines 'First come, first served' and 'Last come, first served' so that, the results derived in earlier sections can be directly used in these.

(a) 'First come, first served': An arriving customer at time t will have to wait till all the customers ahead of him finish their service to enter the counter. Noting that the distribution of the length of service of n customers is the n-fold convolution of the service time distribution, we have

Theorem 1.8

$$d_x \Pr\{W(t) \leq x \,|\, u < W(0) < u + du\} = \sum_{i=0}^{\infty} \sum_{j=1}^{\infty} \int_{y=0}^{x} d_y P_{ij}(t,y) dB_i(u) dB_j(x-y)$$

$$(x > 0) \qquad (1.86)$$

$$\Pr\{W(t) = 0 \,|\, u < W(0) < u + du\} = \sum_{i=0}^{\infty} dB_i(u) P_{io}(t) \qquad (1.87)$$

where $P_{ij}(t,y)$ and $P_{io}(t)$ are given by (1.85) and (1.51) respectively.

(b) 'Last come, first served': With this queue discipline, the waiting time depends on the number of arrivals after time t. Let $V(y,t)$ be the probability distribution of the unexpended service time defined as in (1.78) and (1.79). Let r be the number of arrivals in time y. The customer arriving at t will enter service at the end of the busy period initiated by the number r of customers arrived during the unexpended service time at time t. Thus we have

Theorem 1.9

$$d_x \Pr\{W(t) \leq x\} = \sum_{r=0}^{\infty} \int_{y=0}^{x} d_y V(y,t) e^{-\lambda y} \sum_{k=0}^{r} \frac{(\lambda y)^k}{k!} b_r^{(k)} dG_r(x-y) \qquad (1.88)$$

where

$$dG_r(t) = \sum_{n=r}^{\infty} dG_r^{(n)}(t) \qquad (1.89)$$

$dG_r^{(n)}(t)$ being as given in Theorem 1.1.

1.7 The Waiting Time W(t): An independent study

The waiting time process W(t) in the queue with 'first come,
first served' discipline has been studied extensively by several authors
[see Takács (1964) and the references cited by him). Of particular
interest is the representation.

$$W(t) = \max \{ \sup_{0 \leq \tau \leq t} [X(\tau) - \tau] , W(0) + X(t) - t\} \qquad (1.90)$$

where X(t) is the total service time of the customers arriving in
time (0,t) [Gani and Pyke (1960)]. Most of the studies of W(t)
have been through its transforms and if one is interested in explicit
results, the transforms have to be inverted as has been done by Prabhu
(1960a). It is our intention here to give a direct method which leads
to the explicit expressions for the main transition distribution
functions of W(t). We do this by studying a related process

$$Y^*(t) = u + t - X(t). \qquad (u \geq 0) \qquad (1.91)$$

The stochastic process Y*(t):

Let $\{X(t) , 0 \leq t \leq T\}$ be a separable stochastic process with
non-negative increments, increasing only in jumps and continuous on the
right. Let N(t) be the number of such increases in X(t). Further,
let these jumps occur in a Poisson process and the magnitude of the
jumps have a distribution dB(x) $(0 < x < \infty)$ such that, the joint
distribution of X(t) and N(t) is given by

$$K_n(x,t) = \Pr\{X(t) \le x \; , \; N(t) = n\}$$

$$= e^{-\lambda t} \frac{(\lambda t)^n}{n!} B_n(x) \qquad (1.92)$$

where $B_n(x)$ is the n-fold convolution of $B(x)$ with itself and $B_o(x) = 0$ if $x < 0$ and $= 1$ if $x \ge 0$. We shall also assume that $X(0) = 0$.

Related to $X(t)$, define the process $Y*(t)$ given by (1.91) and the random variable

$$T(u) = \inf\{t \,|\, Y*(t) < 0\} \qquad , \; Y*(0) = u \qquad (1.93)$$

and let

$$G_n(u;x,t) = \Pr\{Y*(t) \ge x \; , \; T(u) > t \; , \; N(t) = n\}$$

$$(u \ge 0 \; , \; x > 0 \quad \text{and} \quad n \ge 0). \qquad (1.94)$$

Lemma 1.3: We have

$$d_x G_n(0;x,t) = e^{-\lambda t} \frac{\lambda^n}{n!} t^{n-1} \, x \, d_x B_n(t-x) \qquad (1.95)$$

Proof: According to the definition of $Y*(t)$, it increases at a unit rate as long as the value of $X(t)$ does not change; but when it changes $Y*(t)$ decreases by an amount equivalent to the size of the jump, which is a random variable with distribution $B(x)$. When $T(0) > t$, the following two figures indicate the changes in $Y*(t)$ for $N(t) = 0$ and $N(t) = n \; (> \; 0)$ respectively.

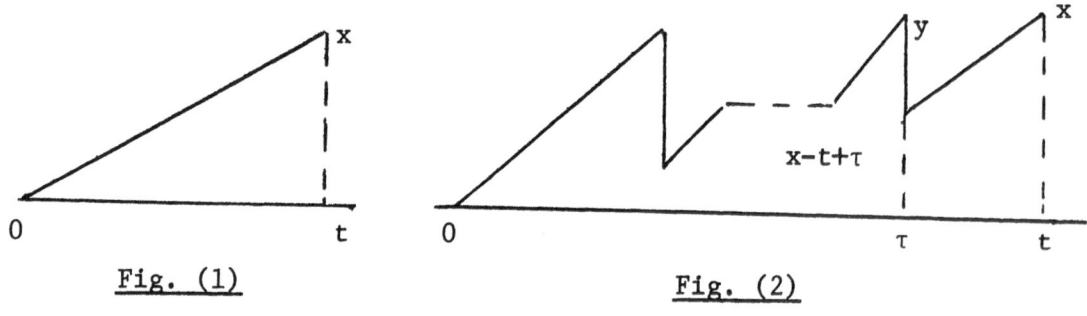

Fig. (1)   Fig. (2)

Suppose there has not been any change in $X(t)$ during $(0,t)$; we have

$$d_x G_0(0;x,t) = e^{-\lambda t} d_x B_0(t-x) \qquad (1.96)$$

[Fig. (1)]. Otherwise, let the last jump occur at $\tau(0 \leq \tau < t)$; the probability of this event is $\lambda d\tau$. Writing down the transitions occurring before and after τ, we have the recurrence relations

$$d_x G_n(0;x,t) = \int_{y=0}^{t-x} \int_{\tau=t-x}^{t} d_x G_{n-1}(0;x-t+\tau+y,\tau)\lambda d\tau \; e^{-\lambda(t-\tau)} dB(y)$$

$$(n \geq 1) \qquad (1.97)$$

[Fig. (2)]. The result (1.95) is obtained by induction. Assuming it to be true for $n = 0,1,2,\ldots,k-1$ and substituting in (1.97) we get

$$d_x G_k(0;x,t) = e^{-\lambda t} \frac{\lambda^k}{(k-1)!} \int_{y=0}^{t-x} d_x B_{k-1}(t-x-y) dB(y) \int_{\tau=t-x}^{t} \tau^{k-2}(x-t+\tau+y) d\tau$$

$$= e^{-\lambda t} \frac{\lambda^k}{(k-1)!} \int_{y=0}^{t-x} \{ \frac{t^k-(t-x)^k}{k} + \frac{t^{k-1}-(t-x)^{k-1}}{k-1} [y-(t-x)] \}$$

$$d_x B_{k-1}(t-x-y) dB(y)$$

$$= e^{-\lambda t} \frac{\lambda^k}{(k-1)!} \{ \frac{t^k-(t-x)^k}{k} + \frac{t^{k-1}-(t-x)^{k-1}}{k-1} [\frac{t-x}{k} - (t-x)] \}$$

$$d_x B_k(t-x) \qquad (1.98)$$

which on simplification gives (1.95). Here we have used the identity (1.6). The result can be extended for $Y^*(0) = u > 0$ as follows.

Lemma 1.4: For $u > 0$, $x > 0$ and $n \geq 1$

$$G_n(u;x,t) = K_n(t+u-x,t) - \sum_{m=1}^{n} \int_0^t d_\tau K_m(\tau+u,\tau) G_{n-m}(0;x,t-\tau). \qquad (1.99)$$

<u>Proof</u>: From the definition (1.94) we can write

$$G_n(u;x,t) = Pr\{Y*(t) \geq x \; ; \; N(t) = n\}$$

$$- Pr\{Y*(t) \geq x \; ; \; T(u) < t; \; N(t) = n\}$$

$$= \Delta_1 - \Delta_2 \; , \; (say) \tag{1.100}$$

in which

$$\Delta_1 = Pr\{u+t-X(t) \geq x \; , \; N(t) = n\}$$

$$= K_n(t+u-x,t). \tag{1.101}$$

To derive Δ_2, consider $\tau(0 \leq \tau < t)$ at which $Y*(\tau) = 0$ for the last time; now, τ is such that $u + \tau < X(\tau) < u + \tau + d\tau$. We therefore have

$$\Delta_2 = \sum_{m=1}^{n} {}_0\!\int^t Pr\{u + \tau < X(\tau) < u + \tau + d\tau, N(\tau) = m\}.$$

$$Pr\{Y*(t) \geq x \; , \; T(u) < t \; , \; N(t) = n | T(Y*(\tau)) > t-\tau \; , \; Y*(\tau)=0, N(\tau)=m$$

$$= \sum_{m=1}^{n} {}_0\!\int^t d_\tau K_m(\tau+u,\tau) \, Pr\{Y*(t-\tau) \geq x \; , \; T(0) > t - \tau \; ,$$

$$N(t-\tau) = n - m\}$$

$$= \sum_{m=1}^{n} {}_0\!\int^t d_\tau K_m(\tau+u,\tau) G_{n-m}(0;x,t-\tau) \tag{1.102}$$

which with Δ_1, gives the required probability.

Now, going back to the definitions of (1.91) and (1.94) of $Y*(t)$ and $G_n(u;x,t)$ respectively, we can write

$$G_n(u;x,t) = Pr\{Y*(t) \geq x \; ; \; T(u) > t; \; N(t) = n | Y(0) = u\}$$

$$= Pr\{u+t-X(t) \geq x \; , \; u+\tau-X(\tau) \geq 0 \quad (0 \leq \tau \leq t); N(t) = n\}$$

$$= Pr\{ \inf_{0 \leq \tau \leq t} [u+\tau-X(\tau)] \geq 0 \; , \; u+t-X(t) \geq x; N(t) = n\}$$

$$= \Pr \{ \sup_{0 \le \tau \le t} \; [X(\tau) - \tau] \le u \; , \; x + X(t) - t \le u; N(t) = n\}$$

$$= \Pr\{W(t) \le u; \; N(t) = n \,|\, W(0) = x\} \tag{1.103}$$

where we have used the expression (1.90) for $W(t)$ [see Prabhu (1965)]

page 103]. Thus we have the explicit expressions given by

Theorem 1.12

Let $N(t)$ be the number of arrivals in $(0,t)$. For, $x > 0$, $u \ge 0$

and $n \ge 0$.

a) $\Pr\{W(t) = 0 \; , \; N(t) = n \,|\, W(0) = x\} = {}_x\!\int^{\infty} d_y G_n(0;y,t)$

$$= e^{-\lambda t} \frac{\lambda^n}{n!} \, t^n \, {}_x\!\int^{\infty} y \, d_y B_n(t-y) \tag{1.106}$$

b) $\Pr\{W(t) \le u \; , \; N(t) = n \,|\, W(0) = x\}$

$$= K_n(t+u-x,t) - \sum_{m=1}^{n} {}_0\!\int^{t} d_\tau K_m(\tau+u,\tau) G_n(0;x,t-\tau). \tag{1.105}$$

[Prabhu (1960)]

Clearly these results follow from (1.95), (1.99) and (1.103).

1.8 The queue M/G/1 with balking [Takács (1961)]:

For convenience consider the queue M/G/1 with unit arrivals.

Customers arrive in a Poisson process and join the queue with

probability 1 if the server is free and with probability $p(<1)$

otherwise. As long as the server is busy the effective arrivals form

a Poisson process with mean λpt. For, consider $A(t)$, the number

of effective arrivals in time $(0,t)$; we have

$$\Pr\{A(t) = r \} = \sum_{s=r}^{\infty} e^{-\lambda t} \frac{(\lambda t)^s}{s!} \binom{s}{r} p^r (1-p)^{s-r} = e^{-\lambda pt} \frac{(\lambda pt)^r}{r!} \; .$$
$$\tag{1.106}$$

Therfore we get

$$dG_i^{(n)}(t) = e^{-\lambda pt} \frac{(\lambda pt)^{n-i}}{(n-i)!} \frac{i}{n} dB_n(t). \qquad (1.107)$$

Similarly, in the Lemma preceding Theorem 1.3, we consider the Poisson

process with mean λpt and obtain the theorem with λ replaced by

λp. However, for $P_{io}(t)$ some more modifications are needed. For,

$$U_i(t) = \sum_{n=0}^{\infty} P\{S_n^{(i)} \le t\} \quad \text{with} \quad T_i \quad \text{as} \quad X_o \quad \text{of the renewal}$$

process $S_n^{(i)}$ is given by

$$dU_i(t) = dG_i(t) + \sum_{n=1}^{\infty} \int_0^t dG_{i+n}(\tau) \, dI_n(t-\tau), \qquad (1.108)$$

where $dG_i(t)$ is given by (1.107) after summing over n and

$$dI_n(t) = e^{-\lambda t} \frac{(\lambda t)^{n-1}}{(n-1)!} dt.$$

The second term in (1.108) can be simplified as

$$\sum_{n=i+1}^{\infty} \frac{\lambda}{n}^{n-i} e^{-\lambda t} dt \int_0^t e^{\lambda(1-p)\tau} dB_n(\tau) \sum_{r=1}^{n-i} (i+r) \frac{(t-\tau)^{r-1}(p\tau)^{n-i-r}}{(r-1)!(n-i-r)!}$$

$$= \sum_{n=i+1}^{\infty} \frac{\lambda^{n-i}}{(n-i-1)!} e^{-\lambda t} dt \int_0^t e^{\lambda(1-p)\tau} [t-(1-p)\tau]^{n-i-2}[t-\tau+\frac{i+1}{n}p\tau]$$

$$dB_n(\tau). \qquad (1.109)$$

Thus we get

$$P_{i0}(t) = {}_0\!\int^t dU_i(\tau)e^{-\lambda(t-\tau)}$$

$$= \sum_{n=i}^{\infty} \frac{i}{n} \frac{\lambda^{n-i}}{(n-i)!} e^{-\lambda t} {}_0\!\int^t e^{\lambda(1-p)\tau}(p\tau)^{n-i} dB_n(\tau)$$

$$+ \sum_{n=i+1}^{\infty} \frac{\lambda^{n-i}}{(n-i-1)!} e^{-\lambda t} {}_{s=0}\!\int^t e^{\lambda(1-p)s} dB_n(s) \; {}_{\tau=s}\!\int^t$$

$$(\tau-s+ps)^{n-i-2}[\tau-s + \frac{i+1}{n}ps]d\tau. \qquad (1.110)$$

The second term in (1.110) can be further simplified as

$$\sum_{n=i+1}^{\infty} \frac{\lambda^{n-i}}{(n-i)!} e^{-\lambda t} {}_{s=0}\!\int^t e^{\lambda(1-p)s}[(t-s + \frac{ips}{n})(t-s+ps)^{n-i-1} - \frac{i}{n}(ps)^{n-i}]$$

$$dB_n(s),$$

which when combined with the first term of (1.110) gives

$$P_{i0}(t) = \sum_{n=i}^{\infty} \frac{\lambda^{n-i}}{(n-i)!} e^{-\lambda t} {}_0\!\int^t e^{\lambda(1-p)\tau}(t-\tau + \frac{ipt}{n}) \; (t-\tau+p\tau)^{n-i-1} dB_n(\tau),$$

$$(i \geq 0). \qquad (1.111)$$

The rest of the discussion follows accordingly.

1.9 Special Cases

We shall consider the particular cases M/D/1 and M/E$_k$/1 with unit arrivals.

(a) The queue M/D/1: Let the service time distribution be given by $B(t) = 0$ if $t < b$ and $= 1$ if $t \geq b$. We have

$$dG_i^{(n)}(t) = \begin{cases} \dfrac{i}{n} e^{-\lambda nb} \dfrac{(\lambda nb)^{n-i}}{(n-i)!} & \text{if } t = nb, \; n \geq i \\[2ex] 0 & \text{otherwise} \end{cases} \qquad (1.112)$$

and

$$
{}^{o}P_{ij}^{(n)}(t) = e^{-\lambda t} \frac{(\lambda t)^{n+j-i}}{(n+j-i)!} - e^{-\lambda t} \lambda^{n+j-i} \sum_{m=i}^{n-1} \frac{i}{m} \frac{(mb)^{m-i}}{(m-i)!}
$$

$$
\frac{(t-mb)^{n-m+j}}{(n-m+j)!} \qquad \text{for} \quad n = \left[\frac{t}{b}\right] \qquad (1.113)
$$

where $[x]$ is the largest integer contained in x. Further, we get

$$
P_{i0}(t) = \sum_{n=0}^{\left[\frac{t}{b}\right]-i} e^{-\lambda t} \frac{(\lambda t)^{n-1}}{n!} \lambda(t-nb) , \quad (i \geq 0). \qquad (1.114)
$$

Using these results in (1.57) we obtain

$$
P_{0j}(t) = \sum_{n=0}^{x} \sum_{k=0}^{x-n} \frac{\lambda^{n+k+j}}{n!(k+j-1)!} e^{-\lambda t} \int_{kb}^{kb+b} \tau^{k+j-1}(t-\tau)^{n-1}(t-\tau-nb)d\tau
$$

$$
- \sum_{n=0}^{x} \sum_{m=1}^{x-n-1} \sum_{k=m}^{x-n} \frac{\lambda^{n+k+j}}{n!m!(k-m+j)!} e^{-\lambda t}(mb)^{m-1} \int_{kb}^{kb+b} (\tau-mb)^{k-m+j} \cdot
$$

$$
(t-\tau)^{n-1} (t-\tau-nb)d\tau, \qquad (1.115)
$$

where $x = \left[\frac{t}{b}\right]$.

Finally, for the steady state probabilities, we get

$$
P_{o}^{*} = 1 - \rho
$$

$$
P_{j}^{*} = (1 - \rho) \sum_{n=0}^{\infty} [e^{-n\rho} \frac{(n\rho)^{n+j-1}}{(n+j-1)!} - e^{-n\rho} \frac{(n\rho)^{n+j}}{(n+j)!}] \qquad (1.116)
$$

where $\rho = \lambda b$.

(b) The queue $M/E_{k}/1$: Let the service time distribution be given

by the k-Erlangian,

$$
dB(t) = e^{-k\mu t} \frac{(k\mu t)^{k-1}}{(k-1)!} k\mu dt. \qquad (1.117)
$$

Simplifying the expressions obtained earlier with $dB(t)$ as given above and writing $dG_i^{(n)}(t) = g_i^{(n)}(t)dt$, we have

$$g_i^{(n)}(t) = e^{-(\lambda+k\mu)t} \frac{i\lambda^{n-i}(k\mu)^{nk}}{n(n-i)! \ (nk-1)!} \ t^{k+n-i-1} \quad (n \geq i) \tag{1.118}$$

$$^{o}P_{ij}^{(n)}(t) = e^{-(\lambda+k\mu)t} \sum_{r=nk}^{nk+k-1} \frac{\lambda^{n+j-i}(k\mu)^r}{(n+j-i)! \ r!} \ t^{n+j-i+r}$$

$$- e^{-(\lambda+k\mu)t} \sum_{m=i}^{n-1} \sum_{r=nk-mk}^{nk-mk+k-1} \frac{i \ \lambda^{n+j-i}(k\mu)^{mk+r}}{m(m-i)! \ (mk-1)! \ (n-m+j)! \ r!}$$

$$_0\int^t \tau^{mk+n-i-1}(t-\tau)^{n-m+j+r}d\tau \tag{1.119}$$

$$P_{io}(t) = 1 - e^{-\lambda t} \sum_{n=i}^{\infty} \frac{\lambda^{n-i}t^{n-i-1}}{\mu(n-i-1)!} - e^{-(\lambda+k\mu)t} \sum_{n=i}^{\infty} \sum_{r=0}^{nk-1} \frac{\lambda^{n-i}(k\mu)^r t^{n-i+r}}{r! \ (n-i)!}$$

$$+ e^{-(\lambda+k\mu)t} \sum_{n=i}^{\infty} \sum_{r=0}^{nk} \frac{\lambda^{n-i}(k\mu)^r t^{n-i+r-1}}{\mu(n-i-1)! \ r!} \tag{1.120}$$

Finally, for the steady state probabilities, we have

$$P_o^* = 1 - \rho$$

$$P_j^* = (1-\rho) \ _0\int^{\infty} e^{-(\lambda+k\mu)t} \sum_{n=0}^{\infty} \frac{\lambda^{n+j-1}(k\mu)^{nk}}{(n-j)! \ (nk-1)'} \ t^{nk+n+j-2} \ [n+j-\lambda t]dt. \tag{1.121}$$

Setting $k=1$ in equations (1.117) - (1.121) we obtain the corresponding probabilities for the queue M/M/1.

2. The Queue GI/M/1 with Group Service

2.1 Description and definitions

Customers arrive at time $t_0 (=0)$, t_1, t_2 ... and the inter-arrival times $t_n - t_{n-1}$ (n=1,2...) form a sequence of identically distributed independent random variables with a common distribution function $B(t)$ (t $\geq$ 0) with a finite non-zero mean. Let

$$\psi(\theta) = {}_0\!\int^\infty e^{-\theta t} \, dB(t), \quad \text{Re } (\theta) \geq 0.$$

The customers are served in batches of variable capacity $\{G_n\}$. We assume that the random variables G_n (n=1,2...) are identically distributed and mutually independent and also independent of the queue length at any moment. Let

$$Pr\{G_n = j\} = b_j \quad (j = 1,2...) \tag{2.1}$$

Let $G(z) = \sum_1^\infty b_r z^r$ with $0 < G'(1) < \infty$, $G'(1)$ being the mean of the batch size distribution.

The service times of batches have the negative exponential distribution $\lambda e^{-\lambda t} dt$ (0 < t < ∞). Under these conditions, the relative traffic intensity ρ_2 is given by

$$\rho_2 = [-\lambda G'(1)\psi'(0)]^{-1} \tag{2.2}$$

The service mechanism is such that the service stops only when there is no customer in the system. Otherwise, even when the number waiting is less than the capacity, the server accepts them as soon as he is free; in such cases, the arriving customers join the batch in service till it is full, without affecting the service time.

Let $Q(t)$ be the number of customers present at the system (queue length) at time t, and we shall define $Q(t_n) = Q(t_n-0)$. Also define the random variable

$$T_i = \inf \{t | Q(t) = 0\} , \quad Q(0) = i, \tag{2.3}$$

and the probabilities

$$^o P_{ij}^{*(n)}(t) = \Pr\{Q(t_n)=j , t_n \leq t , T_i > t_n\} \quad (j > 0) \tag{2.4}$$

and

$$^o P_{ij}^{(n)}(t) = \Pr\{Q(t)=j , t_n \leq t < t_{n+1} , T_i > t\} \quad (j > 0) \tag{2.5}$$

In this system we shall first obtain these probabilities for $i=0$, and then extend the results for $i > 0$. It may be noted, the queues M/G/1 and GI/M/1 may be called dual systems and the duality relations existing between structures of the queue length processes of the two systems will be fully exploited in our treatment. In view of this we have maintained the duality even in our definitions.

2.2 The Busy Period T_o

Consider a busy period that begins with the arrival of a customer. For the probabilities $^o P_{oj}^{*(n)}(t)$ we have

Theorem 2.1:

$$d\, ^o P_{oj}^{*(n)}(t) = \frac{i}{n} e^{-\lambda t} \sum_{k=0}^{n-j} \frac{(\lambda t)^k}{k!} b_{n-j}^{(k)} dB_n(t) \quad (n \geq j > 0) \tag{2.6}$$

where $b_r^{(n)}$ and $B_n(t)$ are the n-fold convolutions of the respective probabilities.

Proof: Consider $(0 \leq \tau < t)$, the time at which the first of the j customers present at t_n, arrived. Clearly, this instant is t_{n-j}; taking into account all the mutually exclusive possibilities during the time intervals before and after τ, we have

$$d^o P_{oj}^{*(j)}(t) = e^{-\lambda t} dB_j(t) \qquad (2.7)$$

and

$$d^o P_{oj}^{*(n+j)}(t) = \sum_{r=1}^{n} \sum_{k=1}^{r} {}_0\!\int^t d^o P_{or}^{*(n)}(\tau) \; e^{-\lambda(t-\tau)} \frac{[\lambda(t-\tau)]^k}{k!} \, b_r^{(k)} dB_j(t-\tau)$$

$$(n \geq 1) \qquad (2.8)$$

These relations are essentially the same as equations (1.8) and (1.9) following Theorem 1.1 and therefore the proof of Theorem 2.1 is identical with the proof of the former.

The result (2.6) is restricted to instants at which arrivals occur and the corresponding result for an arbitrary instant of time is $^o P_{oj}^{(n)}(t)$. We have,

Theorem 2.2: For $j \geq 1$,

$$^o P_{oj}^{(n)}(t) = e^{-\lambda t} \sum_{k=0}^{n-j+1} \frac{(\lambda t)^{k-1}}{k!} \lambda \, b_{n-j+1}^{(k)} \, {}_0\!\int^t [t-(1-\frac{j-1}{b}\,\tau][1-B(t-\tau)] dB_n(\tau).$$

$$(2.9)$$

Proof: Let $\tau(0 < \tau \leq t)$ be the last arrival point. Clearly

$$^o P_{oj}^{(n)}(t) = \sum_{r=\max(1,j-1)}^{n} \sum_{k=0}^{r+1-j} {}_0\!\int^t d^o P_{or}^{*(n)}(\tau) e^{-\lambda(t-\tau)} \frac{[\lambda(t-\tau)]^k}{k!}$$

$$b_{r+1-j}^{(k)} [1-B(t-\tau)] \qquad (2.10)$$

Let $j \geq 2$; substituting from (2.6) and rearranging

$$
{}^{o}P_{oj}^{(n)}(t) = e^{-\lambda t} \sum_{k=0}^{n+1-j} \sum_{s=0}^{n-k-j+1} \int_{0}^{t} \frac{1}{n} \frac{(\lambda \tau)^{s}}{s!} \frac{[\lambda(t-\tau)]^{k}}{k!} \ dB_{n}(\tau)[1-B(t-\tau)]
$$

$$
\sum_{r=k+j-1}^{n-s} r \ b_{n-r}^{(s)} \ b_{r+1-j}^{(k)}
$$

$$
= e^{-\lambda t} \frac{1}{n} \sum_{s=0}^{n+1-j} \sum_{k=0}^{s} b_{n-j+1}^{(s)} \lambda^{s} \int_{0}^{t} \frac{(t-\tau)^{k}}{k!} \frac{\tau^{s-k}}{(s-k)!} \ [\frac{nk+(j-1)(s-k)}{s}]
$$

$$
dB_{n}(\tau)[1 - B(t-\tau)].
$$

$$
= e^{-\lambda t} \frac{1}{n} \sum_{s=0}^{n-j+1} b_{n-j+1}^{(s)} \lambda^{s} \int_{0}^{t} [\frac{(n-j+1)(t-\tau)t^{s-1}}{s!} + \frac{(j-1)t^{s}}{s!}]
$$

$$
dB_{n}(\tau)[1 - B(t-\tau)] \qquad\qquad (2.11)
$$

which gives (2.9). A similar simplification for $j=1$ shows that
(2.9) is true in this case also.

The distribution of the busy period follows directly from Theorem 2.2.
Let N be the number of customers served in a busy period and
$g^{(n)}(t)dt$ be the distribution defined as

$$
g^{(n)}(t)dt = \Pr\{t < T_{o} < t + dt \ ; \ N = n\} \ ; \qquad\qquad (2.12)
$$

This is the joint distribution of the number of customers served in
and the length of a busy period. We have,

Theorem 2.3:
$$
g^{(n)}(t) = e^{-\lambda t} \sum_{s=1}^{n-1} \sum_{r=1}^{n-s} \frac{\lambda^{s+1} t^{s-1}}{s!} [b_{n-r}^{(s)} - b_{n-r}^{(s+1)}] \int_{0}^{t}
$$

$$
[t - (1- \frac{r-1}{n-1} \tau] \ [1-B(t-\tau)]dB_{n-1}(\tau)
$$

$$
- e^{-\lambda t} \sum_{s=1}^{n-1} \sum_{r=1}^{n-s} \frac{\lambda^{s+1} t^{s-1}}{(s+1)!} \frac{n-r}{n-1} b_{n-r}^{(s+1)} \int_{0}^{t} \tau[1-B(t-\tau)]dB_{n-1}(\tau).
$$

$$
(2.13)
$$

Proof: For the distribution $g^{(n)}(t)$, we can write

$$g^{(1)}(t)dt = \lambda e^{-\lambda t}dt[1-B(t)]$$

and

$$g^{(n)}(t)dt = \sum_{r=1}^{n-1} {}^{o}P_{or}^{(n-1)}(t) \; \lambda dt \sum_{k=r}^{\infty} b_k \qquad (2.14)$$

The theorem follows on substituting from (2.9).

In particular, when the customers are served one at a time, this expression simplifies to

$$g^{(n)}(t) = e^{-\lambda t} \frac{\lambda^n t^{n-2}}{(n-1)!} \int_0^t (t-\tau) \; [1-B(t-\tau)]dB_{n-1}(\tau). \qquad (2.15)$$

[Takács (1960)]

2.3 The Busy Period T_i

In this section we shall discuss the probabilities ${}^{o}P_{ij}^{*(n)}(t)$ and ${}^{o}P_{ij}^{(n)}(t)$ for $i,j > 0$ defined in (2.4) and (2.5) respectively and obtain the distribution of the busy period initiated by i customers.

Consider the renewal process $\{U_n\}$, $U_n = u_1+u_2\ldots+u_n$, where $u_r(r=1,2\ldots n)$ have the distribution $dB(t)$ $(0 < t < \infty)$. Let $N(t) = \max\{n|U_n \leq t\}$, so that

$$\Pr\{N(t) = n\} = B_n(t) - B_{n+1}(t). \qquad (n \geq 0) \qquad (2.16)$$

Let $D(t)$ be a compound Poisson process with parameter λ such that

$$\Pr\{D(t) = n\} = \sum_{k=1}^{n} e^{-\lambda t} \frac{(\lambda t)^k}{k!} b_n^{(k)}. \qquad (2.17)$$

It is clear that the random variable given in (2.3) can now be defined with reference to the processes $N(t)$ and $D(t)$ as,

$$T_i = \inf\{t \mid i + N(t) - D(t) \le 0\} .$$ (2.18)

The process $i + N(t) - D(t)$ is non-Markovian; however, the points at which $i + N(\tau) - D(\tau) = 0$ (for some τ) are its points of regeneration. For this process we give below two lemmas which are proved by methods similar to the one adopted in the proof of Lemma 1.1. It should be noted that we should consider the last point τ at which $i + N(\tau) - D(\tau) = 0$ in the proof in place of the first point τ at which $i + A(\tau) - D(\tau) = 0$ in Lemma 1.1.

Lemma 2.1:

$$d_t \Pr\{i + N(t_n) - D(t_n) = j , t_n \le t , T_i > t_n\}$$

$$= e^{-\lambda t} \sum_{k=0}^{n+i-j} \frac{(\lambda t)^k}{k!} b_{n+i-j}^{(k)} dB_n(t)$$

$$- \sum_{m=1}^{n-j} \sum_{k=1}^{m+i} \int_0^t e^{-\lambda \tau} \frac{(\lambda \tau)^k}{k!} b_{m+i}^{(k)} dB_n(\tau)$$

$$d_t \Pr\{N(t_n - t_m) - D(t_n - t_m) = j , t_n - t_m \le t - \tau, T_o > t_n - t_m\}.$$ (2.19)

Lemma 2.2:

$$\Pr\{i + N(t) - D(t) = j , T_i > t , N(t) = n\}$$

$$= e^{-\lambda t} \sum_{k=0}^{n+i-j+1} \frac{(\lambda t)^k}{k!} b_{n+i-j+1}^{(k)} [B_n(t) - B_{n+1}(t)]$$

$$- \sum_{m=1}^{n-j} \sum_{k=1}^{m+i} \int_0^t e^{-\lambda \tau} \frac{(\lambda \tau)^k}{k!} dB_m(\tau) \Pr\{N(t-\tau) - D(t-\tau) = j , T_o > t-\tau,$$

$$N(t-\tau) = n - m\}.$$ (2.20)

Referring to the $Q(t)$ process, we find that the probabilities given by (2.19) and (2.20) correspond to $d^oP_{ij}^{*(n)}(t)$ and $^oP_{ij}^{(n)}(t)$ respectively. Also, we have

$$\Pr\{N(t_n) - D(t_n) = j, \; t_n \le t, \; T_o > t_n\} = {}^oP_{oj}^{*(n)}(t) \qquad (2.21)$$

and

$$\Pr\{N(t) - D(t) = j, \; T_o > t, \; N(t) = n\} = {}^oP_{oj}^{(n)}(t). \qquad (2.22)$$

We summarize these results as

Theorem 2.4:

$$d^oP_{ij}^{*(n)}(t) = e^{-\lambda t} \sum_{k=0}^{n+i-j} \frac{(\lambda t)^k}{k!} \; b_{n+i-j}^{(k)} \; dB_n(t)$$

$$- \sum_{m=1}^{n-j} \sum_{k=1}^{m+i} \int_0^t e^{-\lambda\tau} \frac{(\lambda\tau)^{m+i}}{(m+i)!} \; b_{m+i}^{(k)} \; dB_m(\tau) \; d^oP_{oj}^{*(n-m)}(t-\tau)$$

$$(i \ge 0, \; j, \; n > 0) \qquad (2.23)$$

Theorem 2.5:

$$^oP_{ij}^{(n)}(t) = e^{-\lambda t} \sum_{k=0}^{n+i-j+1} \frac{(\lambda t)^k}{k!} \; b_{n+i-j+1}^{(k)} \; [B_n(t) - B_{n+1}(t)]$$

$$- \sum_{m=1}^{n-j} \sum_{k=1}^{m+i} \int_0^t e^{-\lambda\tau} \frac{(\lambda\tau)^{m+i}}{(m+i)!} \; b_{m+i}^{(k)} \; dB_m(\tau) \; {}^oP_{oj}^{(n-m)}(t-\tau)$$

$$(i, \; n \ge 0, \; j > 0) \qquad (2.24)$$

Consider the busy period initiated by i waiting customers. Let $g_i^{(n)}(t)$ be the joint distribution of the number of customers served in and the length of such a busy period. We have

Theorem 2.6:

$$g_i^{(n)}(t)dt = \sum_{r=1}^{n-1} {}^0P_{ir}^{(n-i-1)}(t) \, \lambda dt \sum_{k=r}^{\infty} b_r \quad (n > i) \qquad (2.25)$$

The right hand side of (2.25) gives the probability that there have been r customers present at t, the capacity of the group in service being at least equal to r and the service is completed in the interval $(t, t+dt)$. Clearly, when the service is in groups, the expression resulting from (2.25) is complicated. Therefore, we give below the simplified expression when the customers are served one at a time. In this case, we have

$$g_i^{(n)}(t) = e^{-\lambda t} \frac{\lambda^n}{(n-1)!} \, t^{n-1} \, [B_{n-i-1}(t) - B_{n-i}(t)]$$

$$- e^{-\lambda t} \sum_{m=1}^{n-i-2} \frac{\lambda^n}{m!(n-m-1)!} \int_{u=0}^{t} \int_{\tau=0}^{u} \tau^{n-m-1}(t-\tau)^{m-1}(t-u)$$

$$[1-B(t-u)]dB_{n-i-1}(u-\tau)dB_{n-i-1-m}(\tau) \qquad (2.26)$$

2.4 The Busy Cycle:

The time interval between the epochs of commencement of two consecutive busy periods is known as a busy cycle. Let $R_i^{(n)}(t)$ $(i \geq 0)$ be the joint distribution of the length of and the number of arrivals in a busy cycle. This can be defined as

$$R_i^{(n)}(t) = \Pr\{Q(t_n) = 0, t_n \leq t, T_i > t_{n-1}\}. \qquad (2.27)$$

When $Q(0) = i > 0$, the first busy cycle will then have the distribution $R_i^{(n)}(t)$ and the rest $R_0^{(n)}(t)$ $(= R^{(n)}(t))$.

We shall consider the two cases $n=1$ and $n > 1$ separately. When $n=1$, all the i (≥ 0) waiting customers at $t_o = 0$ and the one arriving at this epoch should get served in one inter-arrival period $[t_o, t_1)$, say, in k batches. The probability of this event can be given as

$$dR_i^{(1)}(t) = \sum_{k=1}^{i+1} \int_0^t e^{-\lambda\tau} \frac{(\lambda\tau)^{k-1}}{(k-1)!} \lambda d\tau \sum_{\nu=0}^{i} b_\nu^{(k-1)} \sum_{\ell=i+1-\nu}^{\infty} b_\ell \, dB(t)$$

$$= \sum_{k=1}^{i+1} \sum_{\nu=k-1}^{i} [b_\nu^{(k-1)} - b_\nu^{(k)}] \int_0^t e^{-\lambda\tau} \frac{(\lambda\tau)^{k-1}}{(k-1)!} \lambda d\tau dB(t).$$

$$(i \geq 0) \qquad (2.28)$$

When $n > 1$, consider the time $\tau(0 < \tau = t_{n-1} < t)$ at which the last arrival takes place in a busy cycle; let $r(=1, 2,...n-1+i)$ be the number of customers waiting just before this arrival. In the remaining interval $[\tau, t)$, $r+1$ customers get served, say, in k batches. Writing down the probability of this event, we get

$$dR_i^{(n)}(t) = \sum_{r=1}^{n+i-1} \int_0^t d^o P_{ir}^{*(n-1)}(\tau) \, dB(t-\tau)$$

$$\int_{u=0}^{t-\tau} \sum_{k=1}^{r+1} e^{-\lambda u} \frac{(\lambda u)^{k-1}}{(k-1)!} \lambda du \sum_{\nu=0}^{r} b_\nu^{(k-1)} \sum_{\ell=r+1-\nu}^{\infty} b_\ell$$

$$= \sum_{r=1}^{n+i-1} \sum_{k=1}^{r+1} \sum_{\nu=k-1}^{r} [b_\nu^{(k-1)} - b_\nu^{(k)}]$$

$$\int_{\tau=0}^{t} \int_{u=0}^{t-\tau} e^{-\lambda u} \frac{(\lambda u)^{k-1}}{(k-1)!} \lambda du \, d^o P_{ir}^{*(n-1)}(\tau) \, dB(t-\tau). \quad (n > 1)$$

$$(2.29)$$

where $d^o P_{ij}^{*(n)}(t)$ is given by (2.23) (or (2.6) if $i = 0$).

These expressions have elegant forms when the service is not in groups. In view of this, in the rest of the discussion we shall restrict ourselves to the ordinary system GI/M/1 where the service is given to one customer at a time. For the busy cycle distribution $R^{(n)}(t)$, we have

Theorem 2.7:

$$dR^{(1)}(t) = [1-e^{-\lambda t}] \, dB(t) \tag{2.30}$$

$$dR^{(n+1)}(t) = \sum_{r=1}^{n} \frac{r}{n} \int_0^t e^{-\lambda\tau} \frac{(\lambda\tau)^{n-r}}{(n-r)!} \, dB_n(\tau) \, dB(t-\tau)$$

$$- \sum_{r=1}^{n} \frac{r+1}{n+1} e^{-\lambda t} \frac{(\lambda t)^{n-r}}{(n-r)!} \, dB_{n+1}(t) \, , \, (n \geq 1). \tag{2.31}$$

Proof: Equation (2.30) is obvious. For (2.31) we note that

$$dR^{(n+1)}(t) = \sum_{j=1}^{n} \int_{\tau=0}^{t} d^o P_{oj}^{*(n)}(\tau) \, dB(t-\tau) \int_{s=0}^{t-\tau} e^{-\lambda s} \frac{(\lambda s)^j}{j!} \lambda ds. \tag{2.32}$$

Substituting from (2.6) we get

$$dR^{(n+1)}(t) = \frac{\lambda^{n+1}}{n} \int_0^t e^{-\lambda\tau} \, dB_n(\tau) dB(t-\tau) \int_{s=0}^{t-\tau} e^{-\lambda s} \sum_{j=1}^{n} \frac{\tau^{n-j} s^j}{(n-j)! \, j!} \, ds$$

$$- \frac{\lambda^{n+1}}{n!} \int_0^t dB_n(\tau) dB(t-\tau) \int_{s=\tau}^{t} e^{-\lambda s} s^{n-1} (s-\tau) ds$$

$$= \int_0^t dB_n(\tau) dB(t-\tau) [e^{-\lambda\tau} \sum_{r=1}^{n} \frac{(\lambda\tau)^{n-r}}{(n-r)!} - e^{-\lambda\tau} \sum_{r=1}^{n-1} \frac{(\lambda\tau)^{n-r}}{n(n-r-1)!}$$

$$- e^{-\lambda t} \sum_{r=1}^{n} \frac{(\lambda t)^{n-r}}{(n-r)!} + e^{-\lambda t} \lambda\tau \sum_{r=1}^{n-1} \frac{(\lambda t)^{n-r-1}}{n(n-r-1)!}] \tag{2.33}$$

which on simplification gives (2.31).

For the busy cycle distribution $R_i^{(n)}(t)$ we have

Theorem 2.8:

$$dR_i^{(1)}(t) = {}_0\int^t e^{-\lambda\tau} \frac{(\lambda\tau)^i}{i!} \lambda d\tau \ dB(t) \tag{2.34}$$

$$dR_i^{(n+1)}(t) = {}_{s=0}\int^t {}_{u=0}\int^{t-s} {}_{\tau=0}\int^u \sum_{m=1}^{n-1} \frac{\lambda^{n+i-1} s^{n+i-m}}{m! \ (n+i-m)!}$$

$$u^{m-1}(u-\tau)dB_{n-m}(s)dB_m(\tau) \ dB(t-s-\tau)$$

$$+ {}_{\tau=0}\int \ [e^{-\lambda\tau} \sum_{r=1}^{n+i} \frac{(\lambda\tau)^{n+i-r}}{(n+i-r)!} + \frac{e^{-\lambda t}(\lambda\tau)^{n+i}}{(n+i)!}]dB_n(\tau)dB(t-\tau)$$

$$- \sum_{r=0}^{n+i} e^{-\lambda t} \frac{(\lambda t)^{n+i-r}}{(n+i-r)!} \ dB_{n+1}(t) \ , \ (i \geq 0 \ , \ n \geq 1). \tag{2.35}$$

Proof: Equation (2.34) is obvious. For (2.35) we have

$$dR_i^{(n+1)}(t) = \sum_{j=1}^{n+i} {}_{\tau=0}\int^t d^oP_{ij}^{*(n)}(\tau)dB(t-\tau) \ {}_{s=0}\int^{t-\tau} e^{-\lambda s} \frac{(\lambda s)^j}{j!} \ \lambda ds \tag{2.36}$$

Substituting from (2.23) and simplifying we get

$$dR_i^{(n+1)}(t) = \lambda^{n+i+1} {}_{\tau=0}\int^t e^{-\lambda\tau} dB_n(\tau)dB(t-\tau) \ {}_{s=0}\int^{t-\tau} e^{-\lambda s} \sum_{j=1}^{n+i} \frac{\tau^{n+i-j} s^j}{(n+i-j)! \ j!} \ ds$$

$$+ \lambda^{n+i+1} {}_{\tau=0}\int^t e^{-\lambda\tau} dB(t-\tau) \sum_{m=1}^{n-1} {}_{s=0}\int^\tau \frac{s^{n-m+i}}{(n+i-m)!} \ dB_{n-m}(s)dB_m(\tau-s)$$

$${}_{u=0}\int^{t-\tau} \frac{e^{-\lambda u}}{m} \sum_{j=1}^m j \ \frac{u^j(\tau-s)^{m-j}}{j! \ (m-j)!} \ du$$

$$= \Delta_1 + \Delta_2 \ , \ (say). \tag{2.37}$$

We get

$$\Delta_1 = \frac{\lambda^{n+i+1}}{(n+i)!} \int_{\tau=0}^{t} dB_n(\tau)dB(t-\tau) \int_{s=\tau}^{t} e^{-\lambda s}[s^{n+i} - \tau^{n+i}]ds$$

$$= \int_{\tau=0}^{t} [e^{-\lambda\tau} \sum_{r=1}^{n+i} \frac{(\lambda\tau)^{n+i-r}}{(n+i-r)!} + e^{-\lambda t} \frac{(\lambda\tau)^{n+i}}{(n+i)!}]dB_n(\tau)dB(t-\tau)$$

$$- \sum_{r=0}^{n+i} e^{-\lambda t} \frac{(\lambda t)^{n+i-r}}{(n+i-r)!} dB_{n+1}(t); \qquad (2.38)$$

and

$$\Delta_2 = \lambda^{n+i+1} \int_{\tau=0}^{t} dB(t-\tau) \sum_{m=1}^{n-1} \int_{s=0}^{\tau} \int_{u=\tau}^{t} \frac{s^{n+i-m}}{(s+i-m)!} \frac{(u-s)^{m-1}}{m!} (u-\tau)e^{-\lambda u}$$

$$dB_{n-m}(s)dB_m(\tau-s)du. \qquad (2.39)$$

Combining (2.38) and (2.39) we get (2.35).

For further discussion we need the transforms of $R^{(n)}(t)$ and $R_i^{(n)}(t)$. These can be obtained as follows.

Let

$$\Gamma(\theta,\omega) = \sum_{n=1}^{\infty} \omega^n \int_0^{\infty} e^{-\theta t} dR^{(n)}(t). \quad (\text{Re }(\theta) > 0 , |\omega| \le 1)$$

$$(2.40)$$

We have

$$\Gamma(\theta,\omega) = \int_0^{\infty} e^{-\theta t} (1 - e^{-\lambda t})dB(t)$$

$$+ \omega \int_{t=0}^{\infty} e^{-\theta t} \sum_{j=1}^{\infty} \sum_{n=j}^{\infty} \omega^n \int_{\tau=0}^{t} d^0P_{oj}^{*(n)}(\tau)dB(t-\tau)$$

$$\int_{s=0}^{t-\tau} e^{-\lambda s} \frac{(\lambda s)^j}{j!} \lambda ds$$

$$= \omega[\psi(\theta) - \psi(\theta+\lambda)] + \omega \ {}_{t=0}\!\int^\infty e^{-\theta t} dB(t) \ {}_{s=0}\!\int^t e^{-\lambda s} \sum_{j=1}^\infty \frac{(\lambda \gamma s)^j}{j!} \lambda ds$$

$$= \frac{\omega\psi(\theta) - \gamma(\theta,\omega)}{1 - \gamma(\theta,\omega)} \ , \tag{2.41}$$

where we have used the result

$$\sum_{n=j}^\infty \omega^n \ {}_0\!\int^\infty e^{-\theta t} d^\circ P_{oj}^{*(n)}(t) = [\gamma(\theta,\omega)]^j \tag{2.42}$$

where $\gamma = \gamma(\theta,\omega)$ is the unique root in unit circle $|z| < 1$ of the

equation $z = \omega\psi(\theta+\lambda-\lambda z)$. [For discussion about the root see proof

of Theorem 1.2].

Setting $\omega=1$ we get the transfomr of $R(t) = \sum_{n=1}^\infty R^{(n)}(t)$ as

$$\Gamma(\theta) = {}_0\!\int^\infty e^{-\theta t} dR(t) \qquad (\text{Re } (\theta) > 0)$$

$$= \frac{\psi(\theta) - \gamma(\theta)}{1 - \gamma(\theta)} \tag{2.43}$$

[Prabhu (1965)]

Similarly in obtaining the transform of $R_i(t) = \sum_{n=1}^\infty R_i^{(n)}(t)$ we

use the result

$$\sum_{i=0}^\infty \sum_{n=\max(1,j-1)}^\infty z^i \omega^n \ {}_0\!\int^\infty e^{-\theta t} d^\circ P_{ij}^{*(n)}(t) = \frac{\omega[z^j - \gamma^j]\psi(\theta+\lambda-\lambda z)}{z - \omega\psi(\theta+\lambda-\lambda z)}$$

$$(\text{Re } (\theta) > 0 \ , \ |z| < 1 \ , \ |\omega| \leq 1)$$

$$\tag{2.44}$$

which can be obtained from (2.23). Thus we get after some simplifications,

$$\Gamma(\theta,z) = \sum_{i=0}^{\infty} z^i \int_0^{\infty} e^{-\theta t} dR_i(t) \qquad (\text{Re } (\theta) > 0 \ , \ |z| < 1)$$

$$= \{ \frac{[\psi(\theta) - \psi(\theta+\lambda-\lambda z)]z}{1 - z} - \frac{[\psi(\theta) - \gamma(\theta)]\psi(\theta+\lambda-\lambda z)}{1 - \gamma(\theta)} \}$$

$$[z - \psi(\theta+\lambda-\lambda z)]^{-1}. \qquad (2.45)$$

2.5 General transitions of Q(t)

Suppose T' , T'_2, ... are the epochs of commencement of busy periods and $T'_0 = 0$, the instant at which the process starts with an arrival. These epochs $T'_r (r = 0, 1,...)$ form a set of renewal points of the general process for which busy cycles $Z_r = T'_{r+1} - T'_r$ $(r = 0,1,2...)$ are renewal periods. The distributions of these random variables have already been obtained; we have

$$\Pr\{Z_1 \le t\} = R_i(t) \qquad (2.46)$$

and

$$\Pr\{Z_r \le t\} = R(t) \quad (r = 2,3,...). \qquad (2.47)$$

Consider the renewal process $\{S_n\}$, $S_n = Z_1 + Z_2 +...+ Z_n$ $(n=1,2...)$. Let $X(t)$ be the number of renewals $(T'_1 , T'_2, ...)$ in time t so that $X(t) = \max\{n | S_n \le t\}$. Also, let $U_i(t) = E[X(t)]$; we have

$$dU_0(t) = \sum_{k=0}^{\infty} dR^{(k)*}(t) \qquad \text{if} \quad Q(0) = 0 \qquad (2.68)$$

$$dU_i(t) = \sum_{k=0}^{\infty} dR_i(t)*dR^{(k)*}(t) \quad \text{if} \quad Q(0) = i \ge 0 \qquad (2.69)$$

Where we have used $*$ to denote convolution and $R^{(k)*}(t)$ is the k-fold convolution of $R(t)$ with itself.

If $Q(0) = 0$, an explicit form for $dU_o(t)$ can be easily obtained. For,

$$U^*_o(\theta) = {}_0\!\!\int^\infty e^{-\theta t}\, dU_o(t) = \frac{1}{1 - \Gamma(\theta)} \qquad (2.50)$$

where $\Gamma(\theta)$ is given by (2.43). Thus we have

$$U^*_o(\theta) = \frac{1 - \gamma(\theta)}{1 - \psi(\theta)} = [1 - \gamma(\theta)] \sum_{n=0}^\infty [\psi(\theta)]^n. \qquad (2.51)$$

Inverting (2.51) and writing $dV(t) = \sum_{n=0}^\infty dB_n(t)$, we get

$$dU_o(t) = dV(t) - {}_0\!\!\int^t d^oP^*_{o1}(\tau)\,dV(t-\tau). \qquad (2.52)$$

[Prabhu (1965)]

We are now in a position to give probabilistic relations for the general transition probabilities. Suppose at time t a busy period is in progress. For the general transition probabilities $P_{ij}(t)$ defined by

$$P_{ij}(t) = \Pr\{Q(t) = j \,|\, Q(0) = i\}.$$

We have

Theorem 2.9:

$$P_{ij}(t) = {}^oP_{ij}(t) + {}_0\!\!\int^t dU_i(\tau)\,{}^oP_{oj}(t-\tau); \quad (i \geq 0, \; j > 0) \qquad (2.53)$$

in particular,

$$P_{oj}(t) = {}_0\!\!\int^t [dV(\tau) - {}_0\!\!\int^\tau d^oP^*_{o1}(s)\,dV(\tau-s)]\,{}^oP_{oj}(t-\tau) \quad (j > 0) \qquad (2.54)$$

Proof: The busy period that is in progress at time t might be the initial one itself or the one that commenced at time τ $(0 < \tau < t)$; taking into account these two possibilities we have (2.53).

Because of the simplifications possible when $Q(0) = 0$ it may be simpler to express $P_{oj}(t)$ as

$$P_{oj}(t) = {}_0\!\int^t dU_o(\tau) \; {}^oP_{oj}(t-\tau) \quad (j > 0). \tag{2.55}$$

Finally, we shall consider the transition probability $P_{io}(t)$, $(i \geq 0)$, that the server is free at time t. We have

Theorem 2.10:

$$P_{io}(t) = [1 - B(t)] \; {}_0\!\int^t e^{-\lambda\tau} \frac{(\lambda\tau)^i}{i!} \lambda d\tau$$

$$+ \sum_{j=1}^{\infty} {}_0\!\int^t dP^*_{ij}(\tau) \, [1 - B(t-\tau)] \; {}_0\!\int^{t-\tau} e^{-\lambda u} \frac{(\lambda u)^j}{j!} \lambda du$$

$$+ {}_0\!\int^t dU_i(\tau) \, [1 - B(t-\tau) \, [1 - e^{-\lambda(t-\tau)}] \, , \; (i \geq 0), \tag{2.56}$$

where

$$P^*_{ij}(t) = {}^oP^*_{ij}(t) + {}_0\!\int^t dU_i(\tau) \; {}^oP^*_{oj}(t-\tau) \, , \; (\, i \geq 0 \, , \, j > 0). \tag{2.57}$$

and

$${}^oP^*_{ij}(t) = \sum_n {}^oP^*_{ij}{}^{(n)}(t).$$

Proof: Starting with $Q(0) = i$, the event $Q(t) = 0$ can occur in the following ways: (i) There is no arrival after $t=0$. (ii) Arrivals occur after $t=0$; let $\tau(0 < \tau < t)$ be the epoch of last arrival and $Q(\tau) = j$ (>0). In the remaining time $t - \tau$, $j + 1$ customers are served and no arrival takes place. And finally (iii) the situation similar to possibility (ii) above, with $Q(\tau) = 0$; the customer arriving at τ will be served in time $t - \tau$ and no arrival will take place. The three terms in (2.56) give respectively the probabilities of these three mutually exclusive and exhaustive possibilities.

When $Q(0) = 0$, the first and the third terms in (2.56) can be combined to write

$$P_{oo}(t) = {}_0\!\int^t dU_o(\tau)\ [1 - B(t-\tau)]\ [1 - e^{-\lambda(t-\tau)}]$$

$$+ \sum_{j=1}^{\infty} {}_{\tau=0}\!\int^t dP^*_{oj}(\tau)\ [1 - B(t-\tau)]\ {}_{u=0}\!\int^{t-\tau} e^{-\lambda u}\ \frac{(\lambda u)^j}{j!}\ \lambda\,du \tag{2.58}$$

where

$$P^*_{oj}(t) = {}_0\!\int^t dU_o(\tau)\,d\,{}^oP^*_{oj}(t-\tau) \qquad (j > 0) \tag{2.59}$$

where $dU_o(t)$ and $d\,{}^oP^*_{oj}(t)$ are given by (2.52) and (2.6) respectively.

We shall obtain here only the transforms of $P_{oj}(t)$ $(j \geq 0)$.

For this we need the transform of ${}^oP_{oj}(t) = \sum_{n=0}^{\infty} {}^oP^{(n)}_{oj}(t)$, given by (2.9) and (2.10). Using (2.10) we can write

$$\sum_{n=j-1}^{\infty} {}_0\!\int^{\infty} e^{-\theta t} \; {}^o P_{oj}^{(n)}(t) = \sum_{n=j-1}^{\infty} \sum_{r=j-1}^{n} {}_0\!\int^{\infty} e^{-\theta t} \; {}_0\!\int^t d \, {}^o P_{or}^{*(n)}(\tau)$$

$$e^{-\lambda(t-\tau)} \frac{[\lambda(t-\tau)]^{(r+1-j)}}{(r+1-j)!} [1-B(t-\tau)] dt.$$

$$= \frac{[1-\gamma(\theta)][\gamma(\theta)]^{j-1}}{\theta+\lambda-\lambda\gamma(\theta)} \qquad (2.60)$$

after simplification. Combining (2.51) and (2.60) we get

$${}_0\!\int^{\infty} e^{-\theta t} P_{oj}(t) dt = \frac{[1-\gamma(\theta)]^2 [\gamma(\theta)]^{j-1}}{[1-\psi(\theta)](\theta+\lambda-\lambda\gamma(\theta))} \qquad (j > 0) \qquad (2.61)$$

[Conolly (1958)]

By similar arguments from (2.58) and (2.59) we get

$${}_0\!\int^{\infty} e^{-\theta t} P_{oo}(t) dt = \frac{1}{\theta} - \frac{1-\gamma(\theta)}{[1-\psi(\theta)][\theta+\lambda-\lambda\gamma(\theta)]} \qquad (2.62)$$

[Conolly (1958)]

Finally, for the steady state probabilities $P_j^* = \lim\limits_{t\to\infty} P_{oj}(t)$ we have

Theorem 2.11:

$$P_o^* = \begin{cases} 0 & \rho_2 \geq 1 \\[2mm] 1-\rho_2 & \text{if } \rho_2 < 1 \end{cases} \qquad (2.63)$$

$$P_j^* = \begin{cases} 0 & \text{if } \rho_2 \geq 1 \\[2mm] \rho_2 \lambda(1-\zeta) \; {}_0\!\int^{\infty} {}^o P_{oj}(t) dt & \text{if } \rho_2 < 1 \end{cases} \qquad (2.64)$$

where ${}^o P_{oj}(t) = \sum\limits_{n=0}^{\infty} {}^o P_{oj}^{(n)}(t)$ given by (2.9) and ζ is the least

positive root of the equation $z = \psi(\lambda-\lambda z)$ in the unit circle $|z| < 1$.

Alternately (2.64) can also be written as

$$P_j^* = \begin{cases} 0 & \text{if } \rho_2 \geq 1 \\ \\ \rho_2(1-\zeta)\zeta^{j-1} & \text{if } \rho_2 < 1. \end{cases} \qquad (2.65)$$

Proof: Using results from renewal theory,* we can write

$$\lim_{t \to 0} P_{oj}(t) = \frac{1}{E(R)} \int_0^\infty {}^oP_{oj}(t)dt \qquad (2.66)$$

where $E(R)$ is the mean busy cycle. We have

$$E(R) = -\Gamma'(0+)$$
$$= [\lambda \rho_2 (1-\zeta)]^{-1} \qquad (2.67)$$

where we have used $\Gamma(\theta)$ as given by (2.43). Hence we have (2.64). The alternative form (2.65) is obtained by setting $\theta=0$ in (2.60) to get $\int_0^\infty {}^oP_{oj}(t)dt$. The expression (2.63) follows on similar arguments.

2.6: The Waiting Time W(t):

Consider the queue GI/M/1 with the discipline 'first come, first served'. When the service times are negative exponential, because of its Markovian property, deriving the waiting time distribution function is much simpler, if the queue length distribution is known. Thus for a finite t, we can write

$$Pr\{W(t) \leq x | W(0) = 0\} = P_{oo}(t) + \sum_{j=1}^\infty P_{oj}(t) \int_0^x e^{-\lambda y} \frac{(\lambda y)^{j-1}}{(j-1)!} \lambda dy \qquad (2.68)$$

This can be brought to the standard form as $t \to \infty$. Thus we have

---

*See, for instance W. Feller, An Introduction to Probability Theory and its Application, Volume II, John Wiley (1966) Chapter XI.

<u>Theorem 2.12</u>: For $x \geq 0$,

$$\lim_{t \to \infty} Pr\{W(t) \leq x | W(0) = 0\} = 1 - \rho_2 \, e^{-\lambda(1 - \zeta)x} \qquad (2.69)$$

[Smith (1963)]

<u>Proof</u>: Using steady state results from (2.63) and (2.65) on the right hand side of (2.68) we have

$$\lim_{t \to \infty} Pr\{W(t) \leq x | W(0) = 0\} = 1 - \rho_2 + \sum_{j=1}^{\infty} \rho_2(1-\zeta)\zeta^{j-1} \int_0^x e^{-\lambda y} \frac{(\lambda y)^{j-1}}{(j-1)!} \lambda dy$$

$$= 1 - \rho_2 + \lambda\rho_2(1-\zeta) \int_0^x e^{-\lambda(1-\zeta)y} dy$$

$$= 1 - \rho_2 + \rho_2 [1 - e^{-(1-\zeta)x}] \qquad (2.70)$$

which on simplification gives (2.69).

3. Queueing Systems in Discrete Time

In Queueing Theory, systems are studied mostly by considering time as a continuous variable. However, in practice we come across systems in which events occur at discrete 'marks' along the time axis. Examples of such situations may be of electronic installations whose operations are governed by internal clocks, or missile bases which fire at oncoming aircraft at regular intervals etc. Theoretically, the mathematical formalisms required for these discrete time processes and the corresponding continuous time processes are essentially the same; nevertheless, from the practical point of view, it seems worthwhile to point out the major modifications needed in their treatment. We do so in this section and present some discrete time results corresponding to the continuous

time results obtained earlier. It should be noted that the discrete
time distribution corresponding to Poisson is Binomial and
to Negative Exponential is Geometric. Thus we shall call these systems
Geom/G/1 and GI/Geom/1.   For the sake of simplicity we shall restrict
ourselves to the unit arrival and unit service in both these cases.

3.1 The Queue Geom/G/1

Let the discrete time-marks be denoted by $0, \sigma, 2\sigma, 3\sigma, \ldots$ along
the time axis such that the interval between any two consecutive marks
is of length σ. Let the arrivals occur at the epochs $n\sigma - 0$ $(n=0,1,2\ldots)$
with probability $p(=1-q)$; q is the probability of no arrival. The
distribution of the number of arrivals A_n in an interval of length
$n\sigma$, is therefore given by

$$\Pr\{A_n = r\} = \binom{n}{r} p^r q^{n-r} \tag{3.1}$$

The service times $\{v'_k\}$ $(k=1,2\ldots)$ whose commencement and com-
pletion occur at time-marks, are measured in terms of σ. We assume
that these are identically distributed and statistically independent
random varialbes, such that

$$\Pr\{v'_k = r\sigma\} = v_r \quad (r=1,2\ldots) \tag{3.2}$$

Define $Q(t)$ as the number of customers in the system (including
the one at service) at time t and let $Q_n = Q(n\sigma + 0)$ $(n=0,1,2\ldots)$.
Further, we assume that at $t = 0$, the first service starts. Define
the random variable

$$N_i = \min\{n | Q_n = 0\} \quad , \; Q_o = i > 0; \tag{3.3}$$

this denotes the length of the busy period initiated by a queue length i. Let D_n be the number of customers served during an interval of length n and define

$$^oP_{io}^{(k)}(n) = \Pr\{N_i = n \ , \ D_n = k\}. \tag{3.4}$$

We have

Theorem 3.1

$$^oP_{io}^{(k)}(n) = \frac{i}{k} \, p^{k-i} \, q^{n-k+i} \, \binom{n}{k-i} \, v_n^{(k)} \tag{3.5}$$

where $v_n^{(k)}$ is the k-fold convolution of v_n with itself and $v_n^{(0)} = 0$ if $n \neq 0$, $= 1$ if $n = 0$.

Proof: Consider the probability $^oP_{io}^{(k+i)}(n)$, $(k \geq 0)$. Let $m\sigma$ be the time at which the initial i customers complete their service. Accounting for the possible transitions during the intervals before and after $m\sigma$, we have

$$^oP_{io}^{(i)}(n) = q^n \, v_n^{(i)} \tag{3.6}$$

$$^oP_{io}^{(k+i)}(n) = \sum_{m=1}^{n} \sum_{r=1}^{\min(m,k)} \binom{m}{r} \, p^r \, q^{m-r} \, v_m^{(i)} \, ^oP_{ro}^{(k)}(n-m), \quad (k > 0) \tag{3.7}$$

The recurrence relations are discrete analogues of relations (1.8) and (1.9) of section 1. Thus the proof of the present theorem is the discrete version of the inductive proof given there for Theorem 1.1.

It may be noted that the joint distribution of the length of and the number of customers served in a busy period can be obtained from (3.5) by setting i = 1.

Another result which we shall present here is for the probability

$$P_{io}(n) = Pr\{Q_n = 0 | Q_o = i\}. \tag{3.8}$$

In our discussion we need the following Lemma.

Consider the event ε : 'the queueing process changes from a non-null state to a null-state' where we regard an idle period as a null state. This is a (delayed) recurrent event and each of the recurrence times X_n (n=2,3 ...) is composed of an idle period and the following busy period; the initial recurrence time X_1 is the length of the initial busy period (for details see discussion leading to Lemma 1.2 and Theorem 1.4 of section 1). Thus we have

$$S_r = X_1 + X_2 \ldots + X_r \thicksim N_i + N_r + I_r$$

$$\thicksim N_{i+r} + I_r \tag{3.9}$$

where the distribution of N_i is given by (3.5) and the random variable I_r has the negative binomial probability given by

$$Pr\{I_r = n\} = I_r(n) = \binom{n-1}{r-1} p^r q^{n-r}. \tag{3.10}$$

Let

$$U_i(n) = \sum_{r=0}^{\infty} Pr\{S_r = n\} ; \tag{3.11}$$

this is the probability that the event ε occurs at $n\sigma$. We have

Lemma 3.1

$$U_i(n) = \sum_{k=i}^{\infty} p^{k-i} q^{n-k+i} \frac{i}{k} \binom{n}{k-i} v_n^{(k)}$$

$$+ \sum_{k=i+1}^{\infty} p^{k-i} q^{n-k+i} \frac{1}{k-i-1} \binom{n-2}{k-i-2} \sum_{m=1}^{n-1} [(n-1)-(1-\frac{i+1}{k})m] v_m^{(k)}.$$

$$(i > 0) \qquad (3.12)$$

Proof: Using (3.9) we can write

$$U_i(n) = \sum_{k=i}^{\infty} {}^o P_{io}^{(k)}(n) + \sum_{r=1}^{\infty} \sum_{m=1}^{n-1} \sum_{k=i+r}^{\infty} {}^o P_{i+r,o}^{(k)}(m) I_r(n-m). \qquad (3.13)$$

Substituting from (3.5) and (3.10), the second term in (3.13) can be written as

$$\sum_{k=i+1}^{\infty} p^{k-i} q^{n-k+i} \frac{1}{k} \sum_{m=1}^{n-1} v_m^{(k)} \sum_{r=1}^{k-i} (i+r) \binom{m}{k-i-r} \binom{n-m-1}{r-1}$$

$$= \sum_{k=i+1}^{\infty} p^{k-i} q^{n-k+i} \frac{1}{k} \sum_{m=1}^{n-1} v_m^{(k)} [(i+1)\binom{n-1}{k-i-1} + (n-m-1)\binom{n-2}{k-i-2}]$$

$$= \sum_{k=i+1}^{\infty} p^{k-i} q^{n-k+i} \frac{1}{k(k-i-1)} \binom{n-2}{k-i-2} \sum_{m=1}^{n-1} [k(n-1) - m(k-i-1)] v_m^{(k)}$$

$$(3.14)$$

which proves the lemma.

The transition probability $P_{io}(n)$ can now be given as

Theorem 3.2:

$$P_{io}(n) = \sum_{k=i}^{\infty} p^{k-i} q^{n-k+i} \frac{1}{k-i} \binom{n-1}{k-i-1} \sum_{m=1}^{\infty} [n-m(1-\frac{i}{k})] v_m^{(k)}$$

$$(i \geq 0). \qquad (3.15)$$

Proof: Clearly

$$P_{io}(n) = \sum_{m=1}^{n} U_i(m) q^{n-m}.$$

The theorem follows by substituting from (3.12) and simplifying the resulting expression [see, Theorem 1.4 of section 1].

When $Q_o = 0$, even though the details vary, the final result is the same as (3.15) with $i = 0$.

3.2 The Queue GI/Geom 1

Let the customers arrive one at a time with the inter-arrival times v_k $(k=1,2 \ldots)$ having the distribution (3.2). Each customer will be served separately and the service time u_k of a customer has the discrete distribution

$$\Pr\{u_k = j\} = q^{j-1} p \qquad (j = 1,2 \ldots). \tag{3.16}$$

Let Q_n be the number of customers in the system at $n\sigma$ and be defined by $Q_n = Q(n\sigma - 0)$. The random variable

$$N_i = \min\{n | Q_n = 0\} \quad , \quad Q_o = i \tag{3.17}$$

represents the length of a busy period initiated by i customers. Let $t_o = 0$, t_1, $\ldots$ be the time-marks at which the customers arrive. We define

$$^{o}P^{*(k)}_{oj}(n) = \Pr\{Q_n = j , N_o > n , n\sigma = t_k | Q_o = 0\} \quad (k,j > 0) \tag{3.18}$$

which gives the probability of having j customers in the system just before the kth arrival, having avoided an empty queue in the meantime (the arrival at $t_o = 0$ is regarded as the zeroth). Writing down recurrence relations for $^{o}P^{(k)}_{oj}(n)$ (see, proof of Theorem 2.1 of section 2) we have

$$^{o}P^{*(k+j)}_{oj}(n) = \sum_{m=1}^{n} \sum_{r=1}^{\min(m,k)} {}^{o}P^{*(k)}_{or}(n-m) \binom{m}{r} p^r q^{m-r} v_s^{(j)} \quad (k > 0) \tag{3.19}$$

which is identical with (3.7) with obvious changes in notations.   Thus
we have

Theorem 3.3

$$
{}^o P_{oj}^{*(k)}(n) = \frac{j}{k} \, p^{k-j} \, q^{n-k+j} \, \binom{n}{k-j} \, v_n^{(k)}. \tag{3.20}
$$

This result has to be extended to an arbitrary time-mark to be
more useful.   Let

$$
{}^o P_{oj}^{(k)}(n) = \Pr\{Q_n = j \ , \ t_k \le n\sigma < t_{k+1} \ N_o > n | Q_o = 0\}
$$

$$
(k \ge 0 \ , \ j > 0). \tag{3.21}
$$

We have

Theorem 3.4

$$
{}^o P_{oj}^{(k)}(n) = p^{k-j+1} \, q^{n-k+j-1} \, \frac{1}{k-j+1} \, \binom{n-1}{k-j} \sum_{m=0}^{n} \sum_{s=n-m+1}^{\infty} v_m^{(k)} \, v_s \left[n + \left(\frac{j-1}{k} - 1\right)m\right]
$$

$$
(k \ge 0 \ , \ j > 0). \tag{3.22}
$$

Proof: Consider the last arrival epoch.   Accounting for the possible
transitions thereafter, we get

$$
{}^o P_{oj}^{(k)}(n) = \sum_{m=0}^{n} \sum_{r=\max(1,j-1)}^{k} {}^o P_{or}^{*(k)}(m) \, \binom{n-m}{r+1-j} \, p^{r+1-j} \, q^{n-m-r+j-1}
$$

$$
\sum_{s=n-m+1}^{\infty} v_s
$$

$$
= p^{k-j+1} \, q^{n-k+j-1} \, \frac{1}{k} \sum_{m=0}^{n} \sum_{s=n-m+1}^{\infty} v_m^{(k)} \, v_s \sum_{r=j-1}^{k} r \binom{m}{k-r} \binom{n-m}{r+1-j}
$$

$$
= p^{k-j+1} \, q^{n-k+j-1} \, \frac{1}{k(k-j+1)} \, \binom{n-1}{k-j} \sum_{m=0}^{n} \sum_{s=n-m+1}^{\infty} v_m^{(k)} \, v_s
$$

$$
[k(n-m)+m(j-1)] \tag{3.23}
$$

which gives (3.22).

The above theorem leads us to the joint distribution $g_k(n)$ of the number of customers served in and the length of a busy period. We get

Theorem 3.5

$$g_1(n) = q^{n-1} p \sum_{s=n+1}^{\infty} v_s \tag{3.24}$$

$$g_k(n) = p^k q^{n-k} \frac{1}{k-1} \binom{n-2}{k-2} \sum_{m=0}^{n-1} \sum_{s=n-m}^{\infty} (n-m-1) v_m^{(k-1)} v_s , \quad (k \geq 2) \tag{3.25}$$

Proof: Equation (3.24) is obvious. Equation (3.25) follows from the relation

$$g_k(n) = {}^o P_{o1}^{(k-1)} (n-1)p. \tag{3.26}$$

Finally we shall consider the busy cycle, the interval between the commencement of any two successive busy periods. Let $R^{(k)}(n)$ be the joint distribution of the number of arrivals in and the length of a busy cycle. We have

Theorem 3.6

$$R^{(1)}(n) = (1 - q^{n-1}) v_n \tag{3.27}$$

$$R^{(k+1)}(n) = p^{k+1} \frac{1}{k} \sum_{m=1}^{n-1} v_m^{(k)} v_{n-m} \sum_{s=m}^{m+k-1} q^{s-k+1} \binom{s}{k-1} (s+1-m)$$

$$(k \geq 1). \tag{3.28}$$

Proof: When there is only one arrival in a busy cycle we have

$$R^{(1)}(n) = v_n \sum_{j=1}^{n-1} q^{j-1} p ; \tag{3.29}$$

this gives (3.27). Otherwise considering the epoch of **first arrival**
from the last, we can write

$$R^{(k+1)}(n) = \sum_{m=1}^{n-1} \sum_{j=1}^{k} {}^{o}P_{oj}^{*(k)}(m) \; v_{n-m} \sum_{s=j+1}^{n-m} \binom{s-1}{j} p^{j+1} q^{s-j-1}$$

$$(k \geq 1). \qquad\qquad (3.30)$$

The result (3.28) now follows when we substitute for $\;{}^{o}P_{oj}^{*(k)}(m)\;$ from
(3.20).

It should be noted that section 3 has been given only to
demonstrate the equivalence of formalisms in the discrete and continuous
time processes.

BIBLIOGRAPHY

BOOKS

Beneš, V.E. (1963): <u>General Stochastic Processes in the Theory of Queues</u>, Addison Wesley.

Brockmeyer, E., H. L. Halstrøm and A. Jensen (1948): <u>The Life and Works of A. K. Erlang</u>, Aeta Polytechnica Scandinavica (Ap. 287, 1960), The Danish Academy of Technical Sciences.

Cox, D.R. and W. L. Smith (1961): <u>Queues</u>, Methuen & Co., (John Wiley)

Khintchine, A.Y. (1960): <u>Mathematical Methods in the Theory of Queueing</u>, Charles Griffin & Company.

Lee, A. M. (1966): <u>Applied Queueing Theory</u>, St. Martin's Press, (MacMillan).

LeGall, P. (1962): <u>Les Systémes Avec au Sans Attente et Les Processus Stochastiques</u>, Tome 1, Dunod, Paris.

Morse, P. M. (1958): <u>Queues, Inventories and Maintenance</u>, John Wiley & Sons, Inc.

Prabhu, N. U. (1965): <u>Queues and Inventories</u>, John Wiley & Sons.

Riordan, J. (1962): <u>Stochastic Service Systems</u>, John Wiley & SOns.

Saaty, T. L. (1961): <u>Elements of Queueing Theory</u>, McGraw Hill Book Co.

Syski, R. (1960a): <u>Introduction to Congestion Theory in Telephone Systems</u>, Oliver and Boyd (London).

Takács, L. (1962): <u>Introduction to the Theory of Queues</u>, Oxford University Press.
         (1967): <u>Combinatorial Methods in the Theory of Stochastic Processes</u>, John Wiley.

GENERAL

Finch, P. D. (1961): "On the Busy Period in the Queueing System GI/G/1", J. Aust. Math. Soc. 2, pp. 217-227.

Heatheote, C. R. (1964): "Divergent Single Server Queues", Proceedings on the Symposium on Congestion Theory, The University of North Carolina, Chapel Hill, pp. 108-136. (1965)

Keilson, J. and A. Kooharian (1962): "On the General Time Dependent Queue with a Single Server", Ann. Math. Statist 33, pp. 767-791.

Kendall, D. G. (1964): "Some Recent Work and Further Problems in the Theory of Queues", Theory of Probability (Russian) Vol. $\underline{9}$, pp. 1-15.

Kiefer, J. and J. Wolfowitz (1955): "On the Theory of Queues with Many Servers", Trans. Ann. Math. Soc., 78, pp. 1-18.

Kingman, J. F. C. (1962): "The Use of Spitzer's Identity in the Investigation of the Busy Period and Other Quantities in the Queue GI/G/1, J. Aust. Math. Soc. 2, pp. 345-356.

(1964): "The Heavy Traffic Approximation in the Theory of Queues", Proceedings of the Symposium on Congestion Theory, The University of North Carolina, Chapel Hill, pp. 137-169 (1965).

(1966): "On the Algebra of Queues", J. App. Prob. $\underline{3}$, pp. 285-326.

Lindley, D. V. (1952): "The Theory of Queues with a Single Server", Proc. Camb. Phil. Soc. 48, pp. 277-289.

Pollaczek, F. (1964): "Concerning an Analytic Method for the Treatment of Queueing Problems", Proceedings of the Symposium on Congestion Theory, University of North Carolina, Chapel Hill, pp. 1-42. (1965) [Also, references cited at the end of this paper].

Rice, S. O. (1962): "Single Server Systems - II, Busy Periods", Bell Sts. Tech. J. $\underline{41}$, pp. 279-310.

Smith, W. L. (1953): "On the Distribution of Queueing Times", Proc. Camb. Phil. Soc. 49, pp. 449-461.

Spitzer, F. (1956): "A Combinatorial Lemma and its Applications to Probability Theory", Trans. Am. Math. Soc. 82, pp. 323-339.

(1957): "The Wiener-Hopf Equation Whose Kernel is a Probability Density", Duke Math. J. 24, pp. 327-343.

Takács, L. (1963): "The Limiting Distribution of the Virtual Waiting Time and the Queue Size for a Single Server Queue with Recurrent Input and General Service Times", Sankhya, $\underline{A25}$, pp. 91-100.

THE QUEUE M/G/1

Beneš, V.E. (1957): "On Queues with Poisson Arrivals", Ann. Math. Stat. 28, pp. 670-677.

Bhat, U.N. (1966): "On a Stochastic Process Occurring in Queueing Systems M/G/1 and GI/M/1 with Limited Waiting Room", RM 150, Michigan State University, East Lansing, Michigan.

Bhat, U.N. and M. J. Erickson (1966): "An Inventory System as a Queue with Transportation Process", RM 160, Michigan State University, East Lansing, Michigan.

Cox, D.R. (1955): "The Analysis of Non-Markovian Stochastic Processes by the Inclusion of Supplementary Variables", Proc. Camb. Phil. Soc. 51, pp. 433-441.

Finch, P.D. (1960): "On the Transient Behavior of a Simple Queue", J. Roy. Stat. Soc. B22, pp. 277-283.

Gani, J. (1958): "Elementary Methods in an Occupancy Problem of Storage", Math. Ann. 136, pp. 454-465.

Gani, J. and N. U. Prabhu (1959): "The Time Dependent Solution for a Storage Model with Poisson Input", J. Math. and Mech. 8, pp. 653-664.

Gani. J. and N. U. Prabhu (1963): "A Storage Model with Continuous Infinitely Divisible Inputs", Proc. Comb. Phil. Soc. 59, pp. 417-429.

Gani, J. and R. Pyke (1960): "The Content of a Dam as a Supremum of an Infinitely Divisible Process", J. Math. and Mech. 2, pp. 639-652.

Gaver, D.P. Jr. (1959): "Imbedded Markov Chain Analysis of a Waiting Line Process in Continuous Time", Ann. Math. Stat. 30, pp. 698-720.

Hasofer, A.M. (1963): "On the Integrability, Continuity and Differentiability of a Family of Functions Introduced by L. Takács", Ann. Math. Statist. 34, pp. 1065-1069.

Heathcote, C.R. (1961): "On the Queueing Process M/G/1", Ann. Math. Statist. 32, pp. 770-773.

Keilson, J. (1963): "A Gambler's Ruin Type Problem in Queueing Theory", Opns. Res. 11, pp. 570-576.

        (1966): "Some Comments on Single Server Queueing Methods and Some New Results", Proc. Camb. Phil. Soc. 60, pp. 237-251.

Keilson, J. and A. Kooharian (1960): "On Time Dependent Queueing Processes", Ann. Math. Statist. 31, pp. 104-112.

Kendall, D.G. (1951): "Some Problems in the Theory of Queues",
J. Roy. Stat. Soc. B 13, pp. 151-185.

        (1957): "Some Problems in the Theory of Dams",
J. Roy. Stat. Soc. B 19, pp. 207-212.

Kingman, J.F.C. (1963): "On Continuous Time Models in the Theory of
Dams", J. Aust. Math. Soc. $\underline{3}$, pp. 480-687.

Lloyd, E.H. (1963): "The Epochs of Emptiness of a Semi-Infinite Discrete
Reservoir", J. Roy. Stat. Soc. B 25, pp. 131-136.

Luchak, G. (1956): "The Solution of the Single Channel Queueing Equation
Characterized by a Time Dependent Poisson Distributed Arrival Rate
and a General Class of Holding Times", Opns. Res. 4, pp. 711-732.

        (1958): "The Continuous Time Solution of the Equations of
the Single Channel Queue with a General Class of Service Time Distributions by the Method of Generating Function", J. Roy. Stat. Soc.
B. 20, pp. 176-181.

Prabhu, N.U. (1960): "Applications of Storage Theory to Queues with
Poisson Arrivals", Ann. Math. Stat. 31, pp. 475-482.

        (1960): "Some Results for the Queue with Poisson Arrivals",
J. Roy. Stat. Soc. B 22, pp. 104-107.

Prabhu, M.U. and U. N. Bhat (1963): "Further Results for the Queue with
Poisson Arrivals", Opns. Res. 11, pp. 380-386.

        (1963): "Some First Passage Problems and
Their Application to Queues", Sankhya, A 25, pp. 281-292.

Reich, E. (1958), (1959): "On the Integra-differential Equation of
Takács I-II", Ann. Math. Statist, $\underline{29}$, pp. 563-570; $\underline{30}$, pp. 143-148.

Runnenburg, J. Jr. (1966): "On the Use of the Methods of Collective Marks
in Queueing Theory", Proceedings of the Symposium on Congestion Theory,
The University of North Carolina, Chapel Hill, (1965) pp. 399-438.

Takács, L. (1955): "Investigations of Waiting Time Problems by Reduction
to Markov Processes", Acta Math. Acad. Sci. Hung. 6, pp. 101-129.

        (1961): "The Transient Behavior of a Single Server Queueing
Process with a Poisson Input", Proc. Fourth Berkeley Symp. on Math.
Stat. and Prob., Berkeley and Los Angeles, University of California
Press, 2, pp. 535-567.

        (1961) "The Probability Law of the Busy Period for Two Types
of Queueing Processes", Opns. Res. 9, pp. 402-407.

        (1962): "The Time Dependence of a Single Server Queue with
Poisson Input and General Service Times", Ann. Math. Stat. 33,
pp. 1340-1348.

        (1962): "A Generalization of the Ballot Problem and its
Application to the Theory of Queues", J. Amer, Statist, Assoc., $\underline{57}$,
pp. 327-337.

(1966): "Application of Ballot Theorem in the Theory of Queues", Proceedings of the Symposium on Congestion Theory, University of North Carolina, Chapel Hill, (1965), - pp. 337-398. [Also see references cited in this paper].

Yeo, G.F. (1961): "The Time Dependent Solution for an Infinite Dam with Discrete Additive Inputs", J. Roy. Stat. Soc. B 23, pp. 173-179.

(1962): "Single Server Queues with Modified Service Mechanisms", J. Aust. Math. Soc. 2, pp. 499-507.

THE QUEUE GI/M/1

Bhat, U.N. (1965): "On a Stochastic Process Occurring in Queueing Systems", J. App. Prob. $\underline{2}$, pp. 467-469.

(1966): "The Queue GI/M/2 with Service Rate Depending on the Number of Busy Servers", Ann. Inst. Statist. Math., Tokyo, $\underline{18}$, pp. 211-221.

(1967a): "Some Explicit Results for the Queue GI/M/1 with Group Service", Sankhya, $\underline{A29}$, 199-206.

(1967b): "Transient Behavior of Multiserver Queues with Recurrent Input and Exponential Service Times", Journal of App. Prob. (to appear).

Conolly, B.W. (1958): "A Difference Equation Technique Applied to the Simple Queue with Arbitrary Arrival Interval Distributions", J. Roy. Stat. Soc. B 20, pp. 167-175.

(1959): "The Busy Period in Relation to the Queueing Process GI/M/1", Biometrika 46, pp. 246-251.

Kendall, D.G. (1953): "Stochastic Processes Occurring in the Theory of Queues and Their Analysis by the Method of the Imbedded Markov Chain", Ann. Math. Stat. 24, pp. 338-354.

Prabhu, N.U. (1964): "A Waiting Time Process in the Queue GI/M/1", Aeta. Math. Acad. Sci. Hung. $\underline{15}$, pp. 363-371.

Shanbhag, D.N. (1963): "On Queues with Poisson Service Time", Aust. J. Stat. 5, pp. 57-61.

Takács, L. (1960): "Transient Behavior of a Single Server Queueing Process with Recurrent Input and Exponentially Distributed Service Times", Opns. Res. 8, pp. 231-245.

(1961): "The Probability Law of the Busy Period for Two Types of Queueing Processes", Opns. Res. 9, pp. 402-407.

(1962): "A Single Server Queue with Recurrent Input and Exponentially Distributed Service Times", Opns. Res. 10, pp. 395-399.

(1964): "Application of Ballot Theorems in the Theory of Queues", Proceedings of the Symposium on Congestion Theory, University of North Carolina, Chapel Hill, (1965), pp. 337-398.

Wisehart, D.M.G. (1956): "A Queueing Distribution with x^2 Service Time Distribution", Ann. Math. Stat. 27, pp. 768-779.

_____ (1959): "A Queueing System with Service Time Distribution of Mixed Chi-squared Type", Opns. Res. 7, pp. 174-179.

Wu Fang (1962): "Some Results About the Queueing System GI/E/1", Chinese Math. (Engl. Transl.) 1, pp. 205-216.

SOME SPECIAL QUEUES

Bailey, N.T.J. (1954): "A Continuour Time Treatment of a Simple Queue Using Generating Functions", J. Roy. Stat. Soc. B 16, pp. 288-291.

Borel, E. (1942): "Sur l'emploi du théorème de Bernoulli pour faciliter le calcul d'une infinité de coefficients, Application au probleme de l'attente à un guichet", Compt. Rend. Acad. Sci. Paris, 214, pp. 452-456.

Burke, P.J. (1956): "The Output of a Queueing System", Opns. Res. 4, pp. 699-704.

Champernowne, D.G. (1956): "An Elementary Method of the Solution of the Queueing Problem with a Single Server and a Constant Parameter", J. Roy. Stat. Soc. B 18, pp. 125-128.

Clarke, A.B. (1956): "A Waiting Time Process of Markov Type", Ann. Math. Stat. 27, pp. 452-459.

Conolly, B.W. (1958): "A DIfference Equation Technique Applied to the Simple Queue", J. Roy. Stat. Soc. B 20, pp. 165-167.

Crommelin, C.D. (1932): "Delay Probability Formulae When Holding Times are Constant", P. O. Elect. Engrs' J. 25, pp. 41-50.

Karlin, S. and J. McGregor (1958): "Many Server Queueing Processes with Poisson Input and Exponential Service Times", Pacific J. Math. 8, pp. 87-118.

Ledermann, W., and G. E. Reuter (1954): "Spectral Theory for the Differential Equations of Simple Birth and Death Processes", Phil. Trans. Roy. Soc. London, A 246, pp. 321-369.

Morse, P.M. (1955): "Stochastic Properties of Waiting Lines", J. Opns. Res. Soc. Am. 3, pp. 255-261.

Neuts, M.F. (1964): "The Distribution of the Maximum Length of a Poisson Queue During a Busy Period", Opns. Res. 12, pp. 281-285.

Prabhu, N.U. (1962): "Elementary Methods for Some Waiting Time Problems", Opns. Res. 10, pp. 559-566.

Saaty, T.L. (1960): "Time Dependent Solution of Many Server Poisson Queue", Opns. Res. 8, pp. 755-772.

Tanner, J.C. (1961): "A Derivation of the Borel Distribution", Biometrika 48, pp. 222-224.

SOME DISCRETE QUEUES

Bhat, U.N. (1964): "On Single-Server Bulk-Queueing Processes with Binomial Input", Opns. Res. 12, pp. 527-533.

Bondreau, P.E., Griffin, J.S. Jr., and M. Kac (1962): "A Discrete Queueing Problem with Variable Service Times", IBM Jour. of Res. and Development 6, pp. 406-418.

Hirsch, W.H., Conn, J. and C. Siegel (1961): "A Queueing Process with an Absorbing State", Communications on Pure and App. Math. 14, pp. 137-153.

Meisling, T. (1958): "Discrete Time Queueing Theory", Opns. Res. 6, pp. 96-105.

Natarajan, R. (1962): "Discrete Time Bulk Service Queueing Processes", Defence Sci. J., Defence Sci. Lab., Delhi, pp. 318-326.

Offsetdruck: Julius Beltz, Weinheim/Bergstr

Lecture Notes in Operations Research and Mathematical Economics

Ökonometrie und Unternehmens-forschung
Econometrics and Operations Research

Herausgegeben von / Edited by

M. Beckmann, Bonn; R. Henn, Karlsruhe; A. Jaeger, Cincinnati; W. Krelle, Bonn; H. P. Künzi, Zürich; K. Wenke, Ludwigs-hafen; Ph. Wolfe, Santa Monica (Cal.)
Geschäftsführende Herausgeber / Managing Editors
W. Krelle und H. P. Künzi